WHAT YOUR COLLEAGUES ARE SAYING . . .

"Too often, in mathematics, we are in a rush to move from the concrete to abstract. In doing so, we overlook the visual. This book is an excellent reminder of the importance of the visual. It not only helps us understand how visuals can help students see mathematics but also helps us see what students see when they are doing mathematics. This book is a necessity for every K–6 teacher."

Peter Liljedahl
Author, Building Thinking Classrooms
Burnaby, BC, Canada

"*Seeing the Math You Teach* delivers exactly what it promises—math you can actually see. The visuals are clean, clear, and do the heavy lifting, making tricky concepts far more intuitive for both students and teachers. If you've ever struggled to explain fractions without resorting to interpretive dance, this book is for you."

Fawn Nguyen
Director, STEM Initiatives
Oak View, CA

"If we're serious about equity in math, we must change how we teach it. *Seeing the Math You Teach* gives parents and teachers the tools to do just that—by using visual models and brain-friendly strategies that support all kids, not just those labeled as 'high achievers,' in developing real mathematical understanding."

Pam Seda
Co-author of *Choosing to See: A Framework for Equity in the Math Classroom*
Atlanta, GA

"*Seeing the Math You Teach* is an essential resource for educators looking to deepen their mathematical understanding and better support their students. Through clear visuals and a progression of conceptual development, this book helps teachers see how students think and learn, allowing them to meet each learner where they are. A must-have for building instructional confidence and fostering meaningful mathematical understanding in the classroom!"

Graham Fletcher
Math Specialist
Atlanta, GA

"This book is a valuable resource for teachers who want to make math more accessible and meaningful for *all* students. With clear explanations, visual support, and practical strategies, it helps break down abstract concepts so that every learner can engage and build understanding. Grounded in how children learn, this book offers thoughtful guidance for creating lessons that support problem-solving, confidence, and a genuine connection to math."

Melynee Naegele
Instructional Coach, Osage County Interlocal Cooperative
Claremore, OK

"Teachers often ask, 'How can I ratchet up the quality of my math teaching to better engage my students in mathematical thinking and to deepen their understanding of math beyond answer-getting?' Call this helpful book 'A Compendium of Math Representations' because it provides a slew of topic-specific ways to picture mathematics. By including these multiple representations in our lessons, we ensure that all the active, yet different, brains in our classrooms can 'get it.'"

Steven Leinwand
Consultant, Mentor, Change Agent
Washington, DC

"*Seeing the Math You Teach* is an essential resource that transforms mathematics into a visual and accessible experience for all students. This comprehensive guide will undoubtedly become a favorite among teachers, showing signs of frequent use as they refer to it time and time again!"

Ann Elise Record
Ann Elise Record Consulting LLC
Concord, NH

"*Seeing the Math You Teach* is a valuable resource that brings clarity to mathematical representations. It will help you make essential concepts more accessible for all learners. The practical strategies and visual models will empower you to foster deeper understanding and greater confidence in math."

John SanGiovanni
Education Coordinator,
Instructional Facilitator for Elementary Mathematics,
Howard County Public Schools
Westminster, MD

"I've shared about the 'Sweet Spot of the C-R-A model,' where you do an activity that includes all 3 and the power that has with building your students' understanding. *Seeing the Math You Teach* is full of representations to help you get to that 'Sweet Spot.'"

Christina Tondevold
Founder, Build Math Minds
Orofino, ID

"If you want to see math visually, then this is the book for you! *Seeing the Math You Teach* shows us what math looks like. It's an easy-to-read reference book that will be on my desk to share with teachers (and students!)"

Laura Vizdos Tomas
K–5 Math Coach, School District of Palm Beach County
Cofounder, LearningThroughMath.com
West Palm Beach, FL

"With clear explanations and powerful visual representations, *Seeing the Math You Teach* is an indispensable resource for K–6 teachers looking to deepen their understanding of mathematics and how students learn it. Packed with practical strategies, this book empowers educators to guide students beyond rote memorization toward true mathematical comprehension by bridging the gap between abstract concepts and hands-on, visual learning. A must-have for any teacher committed to helping students see, explore, and master math with confidence."

Chase Orton
Author, *The Imperfect and Unfinished Math Teacher*
Culver City, CA

"*Seeing the Math You Teach* is an essential resource for teachers, instructional aides, special educators, and those entering the profession through alternative certification. As a district math coordinator, I appreciate how the book's clear visuals, concise text, and embedded video clips make complex mathematical representations immediately accessible—not just for educators, but also for parents and caregivers supporting math learning at home. An eye-opening approach to visualizing mathematics!"

Tara Fulton
K–8 Math Coordinator, Crane School District
Yuma, AZ

"This book is a must-read for educators and caregivers who want to foster sense-making in mathematics. The authors present clear, comprehensive visual representations that make abstract concepts understandable. A valuable resource for anyone who is eager to provide children with rich and connected mathematical experiences."

Amy Chang
Official Team Member, Building Thinking Classrooms
South Hadley, MA

"*Seeing the Math You Teach* is more than just a resource—it's a toolbox for teachers, packed with practical ideas to help make math visible and meaningful. It gives you simple ways to break down big concepts and help every student truly get it."

Bryan Borden
Instructional Coach, Buckeye Elementary School District
Buckeye, AZ

SEEING the MATH YOU TEACH

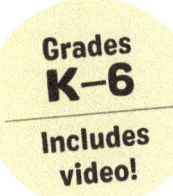

Grades K–6
Includes video!

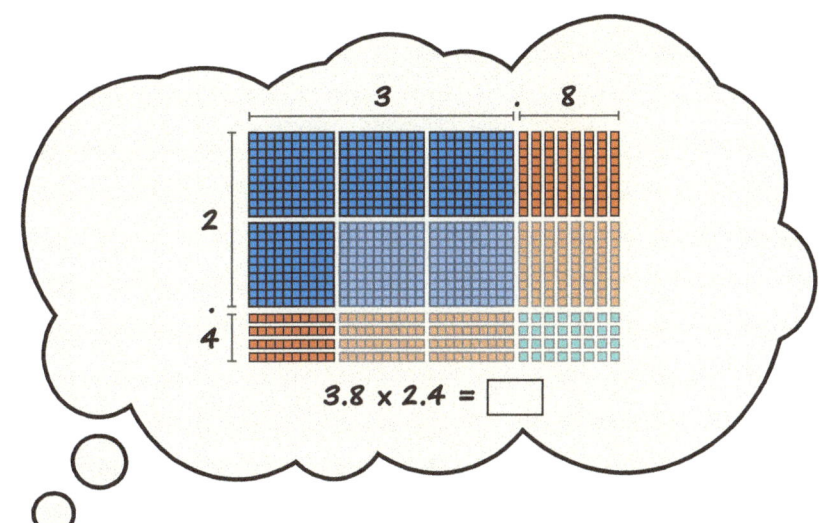

SEEING the MATH YOU TEACH

An Elementary Teacher's Quick-Guide to Meaningful Mathematical STRATEGIES and REPRESENTATIONS

KIMBERLY RIMBEY
KATIE BASHAM
CHRYSTE BERDA

CORWIN

FOR INFORMATION:

Corwin
A SAGE Company
2455 Teller Road
Thousand Oaks, California 91320
(800) 233-9936
www.corwin.com

SAGE Publications Ltd.
1 Oliver's Yard
55 City Road
London EC1Y 1SP
United Kingdom

SAGE Publications India Pvt. Ltd.
Unit No 323-333, Third Floor, F-Block
International Trade Tower Nehru Place
New Delhi 110 019
India

SAGE Publications Asia-Pacific Pte. Ltd.
18 Cross Street #10-10/11/12
China Square Central
Singapore 048423

Vice President and
 Editorial Director: Monica Eckman
Associate Director and
 Publisher, STEM: Erin Null
Senior Editorial Assistant: Nyle De Leon
Production Editor: Tori Mirsadjadi
Copy Editor: Talia Greenberg
Typesetter: C&M Digitals (P) Ltd.
Proofreader: Wendy Jo Dymond
Cover Designer: Gail Buschman
Marketing Manager: Margaret O'Connor

Copyright © 2025 by Corwin Press, Inc.

All rights reserved. Except as permitted by U.S. copyright law, no part of this work may be reproduced or distributed in any form or by any means, or stored in a database or retrieval system, without permission in writing from the publisher.

When forms and sample documents appearing in this work are intended for reproduction, they will be marked as such. Reproduction of their use is authorized for educational use by educators, local school sites, and/or noncommercial or nonprofit entities that have purchased the book.

All third-party trademarks referenced or depicted herein are included solely for the purpose of illustration and are the property of their respective owners. Reference to these trademarks in no way indicates any relationship with, or endorsement by, the trademark owner.

ISBN 978-1-0719-8466-6

DISCLAIMER: This book may direct you to access third-party content via web links, QR codes, or other scannable technologies, which are provided for your reference by the author(s). Corwin makes no guarantee that such third-party content will be available for your use and encourages you to review the terms and conditions of such third-party content. Corwin takes no responsibility and assumes no liability for your use of any third-party content, nor does Corwin approve, sponsor, endorse, verify, or certify such third-party content.

Math Topics

Acknowledgments — xiii
About the Authors — xv

Introduction — 1

Chapter 1: Whole Number and Decimal Place Value — 8
Ones and Tens on a Ten Frame — 10
Whole Number Place Value — 11
Place Value Organization — 14
Decimal Place Value — 16
Decimal Point — 18
Comparing Numbers — 19
Comparing Decimals — 20

Chapter 2: Math Symbols and Properties — 21
Commonly Used Math Symbols — 22
Signs of Comparison — 23
Properties of Addition — 25
Properties of Multiplication — 26
Order of Operations — 27

Chapter 3: Estimation (Including Rounding) — 28
Estimation (Including Rounding) — 30
Rounding Large Numbers — 31
Rounding With Decimals — 32
Estimation Strategies — 33
Estimating With Benchmark Fractions — 34

Chapter 4: Addition and Subtraction Using Place Value — 36
Addition — 38
Subtraction — 38
The Relationship Between Addition and Subtraction — 39
Writing Equations for Addition and Subtraction Word Problems — 40

Addition and Subtraction Word Problems	42
Addition and Subtraction Two-Step Word Problems	43
Adding Whole Numbers	44
Adding Larger Whole Numbers	45
Subtracting Using the Adding-On Method	46
Subtracting Whole Numbers	47
Subtracting Larger Whole Numbers	49
Subtracting Using the Traditional Method	50
Adding and Subtracting Decimals	51

Chapter 5: Multiplication and Division Using Place Value — 54

Multiplication	56
Division	56
The Relationship Between Multiplication and Division	57
Writing Equations for Multiplication and Division Word Problems	58
Multiplication Word Problems	59
Multiplication as a Scale Factor	60
Multiplication Table	61
Single-Digit Multiplication Strategies	62
Multiplying by Multiples of Ten	63
Multi-Digit Multiplication Strategies	64
Understanding the Traditional Multiplication Method	66
Multiplying Decimals	67
Division Word Problems	69
Fractions as Division	70
Division Strategies	72
Understanding the Partial Quotient Division Method	73
Understanding the Traditional Division Method	74
Remainders	75
Dividing Decimals	76
Two-Step Word Problems ($+ - \times \div$)	77

Chapter 6: Special Topics With Whole Numbers — 79

Odd and Even Numbers	80
Hundreds Chart	81
Multiples and Common Multiples	82
Factors	83
Common Factors	84
Prime Numbers	85
Composite Numbers	86
Exponents	87

Chapter 7: Negative and Positive Numbers — 88

- Integers — 90
- Positive and Negative Numbers — 90
- Comparing Integers — 91
- Absolute Value — 92

Chapter 8: Fraction Basics — 94

- Partitioned Shapes — 96
- Fraction Symbols and Representations — 97
- Unit Fractions — 98
- Using Unit Fractions — 98
- Decomposing Fractions — 99
- Fractions for Whole Numbers — 100
- Equivalent Fractions — 101
- Comparing Fractions With the Same Denominator — 102
- Comparing Fractions With the Same Numerator — 102
- Comparing Fractions With Different Denominators and Numerators — 103
- Finding Common Denominators — 104

Chapter 9: Add, Subtract, Multiply, and Divide Fractions — 106

- Adding and Subtracting Fractions — 108
- Adding and Subtracting Mixed Numbers — 109
- Solving Addition and Subtraction Word Problems Involving Fractions — 110
- Solving Multiplication and Division Word Problems Involving Fractions — 111
- Multiplying Fractions by Whole Numbers — 112
- Multiplying Fractions by Fractions — 113
- Multiplying Fractions With Mixed Numbers — 114
- Multiplying Fractions by Fractions Using Decomposition — 115
- Division With Fractions — 116

Chapter 10: Relationships Between Fractions, Decimals, and Percents — 118

- Relationships Between Fractions, Decimals, Percents, and Ratios — 120
- Decimal Fractions — 121
- Converting Fractions to Decimals and Percents — 122

Chapter 11: Ratios and Rates — 124

- Ratios — 126
- Using Ratios as Rates — 127
- Equivalent Ratios — 128

Chapter 12: Algebraic Expressions, Equations, and Inequalities — 132

- Algebra Vocabulary — 134
- Writing Expressions — 135
- Evaluating Expressions — 136
- Generating and Analyzing Patterns — 137
- Writing Equations — 138
- Solving Equations (Two-Step) — 139
- Solving Inequalities — 141

Chapter 13: Coordinate Planes — 143

- Coordinate Geometry — 144
- Coordinate Plane — 145
- Graphing in the Coordinate Plane — 146
- Graphing Numerical Patterns — 147

Chapter 14: Geometry — 148

- Attributes of Shapes — 150
- Geometry Vocabulary — 151
- Symmetry — 152
- Adjacent and Congruent Sides — 152
- Angles — 153
- Angle Measures — 154
- Measuring Angles — 155
- Polygons — 156
- 2-D Shapes — 157
- Classification of Triangles — 158
- Classification of Quadrilaterals — 159
- 3-D Shapes — 161
- Composite Shapes — 163

Chapter 15: Measurement — 164

- Measurement: Length — 166
- Length Measurement: Word Problems — 167
- Length Measurement: U.S. Standard — 169
- Weight Measurement: U.S. Standard — 169
- Liquid Volume Measurement: U.S. Standard — 170
- Length, Weight, and Liquid Volume Measurement: Metric — 171
- Telling Time — 172

Elapsed Time	173
Money	174
Perimeter and Area	175
Perimeter and Area of Rectangles With Fractional Side Lengths	176
Area of a Parallelogram	177
Area of a Triangle	177
Solving Area Problems	178
Surface Area	179
Volume	180

Chapter 16: Data — 182

Data	184
Picture Graphs	185
Bar Graphs	186
Line Plots (Dot Plots)	187
Histograms	188
Systematic Listing and Counting	189
Statistical Variability	190
Measures of Center/What Is Typical?	191
Measures of Variation/The Spread of Data	192
Box Plots (Definitions)	193
Box Plots (Example)	194
Mean Absolute Deviation	195

Topic Index — 197

Note From the Publisher: The authors have provided video and web content throughout the book that is available to you through QR (quick response) codes. To read a QR code, you must have a smartphone or tablet with a camera. We recommend that you download a QR code reader app that is made specifically for your phone or tablet brand.

Acknowledgments

From Kimberly Ann Rimbey:

The evolution of this project spans several years and countless contributors, each leaving an indelible mark on its journey. My heartfelt gratitude goes to the Rodel Foundation of Arizona, founded by Don Budinger, for fostering an environment where innovation and creativity could thrive. Their unwavering commitment to supporting teachers, students, and families in their math journeys laid the foundation for this work. A special thank-you to the incredible Rodel team—Charmaine Bolden and Frieda Pollack, in particular—for their collaboration and dedication throughout the many iterations of what ultimately became *Math Power: Simple Solutions for Mastering Math*.

After a year of designing the prototype for this project, I had the privilege of passing the torch to an exceptional writing team: Chryste Berda, our inspiring team leader, and Katie Basham, our brilliant lead writer. Their tireless efforts, alongside ongoing support from the Rodel team, culminated in a 20,000-book run that enriched teachers, families, and students across Arizona.

Today, I am thrilled to bring you this new version, made possible by the outstanding partnership with the Corwin team. Once again working alongside Katie and Chryste has been an honor, and I am so excited to share this resource with you, your students, and their families.

Finally, to my husband and family—thank you for your unwavering support, boundless encouragement, and love that fuels everything I do. I couldn't do this without you.

From Katie Basham:

My heartfelt thanks to **all** of the hardworking, focused, and compassionate educators whom I have had the pleasure of working alongside. I have learned more about myself as an educator and as an advocate for children because of you—thank you. To Kim for lighting the spark and to Chryste for feeding the flame; I am indebted to you for your insight and perspective.

Math makes sense. I wish that I'd known this more finitely when I was a young, impressionable student. I wish my teachers had known it. Thank you to my parents, who never questioned my capabilities and who, decades later, continue to have confidence that I am capable of anything.

To my husband, Craig, thank you for not only the encouragement but for the unwavering support and love. To William and Sydney, quick to boast that their mother is a twice-published author, thank you. I believe in myself because you all believe in me.

From Chryste Berda:

My sincere appreciation goes to my family and math friends. The unyielding love, support, and encouragement that my husband, Christopher, and my two sons, Christian and Collin, give me daily carried me through this process. The close listening and timely questioning from my mom, Carolyn E. Danielle, a poet and writer who has been my thought-partner for every piece of writing since junior high, brought me clarity through stuck moments. I also want to acknowledge my friends and colleagues Kim Rimbey and Katie Basham—Kim for connecting us with Corwin, originating the idea of this book, and trusting me to lead its creation at the Rodel Foundation of Arizona under the guidance of Don Budinger, who generously funded our team's early work; and Katie for her clarity of elementary math concepts and her ability to make these ideas both visible and understandable to others, which has brought much-needed clarity to our field. I would also like to thank my local public libraries and coffee shops for providing workspaces where my ideas could flow and where my teacup never ran dry. These working spaces got me through many all-day writing sessions.

From All:

Our deepest thanks go to Associate Director and Publisher Erin Null and the incredible Corwin team for their insightful guidance and steadfast support as we navigated this endeavor. Your expertise and encouragement have been invaluable to our writing team, and we are profoundly grateful for your partnership in bringing this project to life.

Publisher's Acknowledgments

Corwin gratefully acknowledges the contributions of the following reviewers:

Susan Dalka
Elementary Curriculum Specialist, Avon Public Schools
Avon, CT

Susie Katt
K–2 Mathematics Coordinator, Lincoln Public Schools
Lincoln, NE

Alison J. Mello
Author, Former Superintendent
Math Consultant, Alison Mello Math Consulting, LLC
North Attleboro, MA

Georgina Rivera
Principal, Charter Oak International Academy
Hartford, CT

About the Authors

Kimberly Rimbey is an author, inventor, entrepreneur, speaker, consultant, coach, mentor, advocate, and, proudest of all, teacher through and through. Kim currently serves as the Chief Learning Officer and CEO at KP® Mathematics and an official Building Thinking Classrooms consultant alongside Peter Liljedahl. A lifelong teacher and learner, her heart's work centers on equipping teachers and helping them fall in love with teaching and learning over and over again.

Always a teacher at heart, Kim has held several leadership positions, including Executive Director of Curriculum and Instruction, Chief Learning Officer, and Mathematica Program Area Coordinator. That said, everything Kim has done in her career is based on what she learned during her eighteen years as a mathematics coach and classroom teacher.

Kim is National Board Certified in Early Adolescent Mathematics, and she is a recipient of the Presidential Award for Excellence in Mathematics Teaching. Kim is the co-inventor of KP® Ten-Frame Tiles and has authored and co-authored several publications, including *Mastering Math Manipulatives* and *Meaningful Small Groups in Math* for Corwin, *Math Power: Simple Solutions for Mastering Math* for the Rodel Foundation of Arizona, and, most recently, *The Amazing Ten Frame* series for KP Mathematics.

Kim earned her bachelor's degree in Elementary Education and Mathematics from Grand Canyon University, her master's degrees in Early Childhood Education and Educational Leadership from Arizona State University and Northern Arizona University, and her PhD in Curriculum and Instruction from Arizona State University. Kim lives in Phoenix, Arizona, where she continues to inspire teachers and their leaders.

Katie Basham is an aggressive Scrabble player, avid reader, and an affectionate mother and wife. She is also an instructional specialist in the Pacific Northwest, where she leads professional development and works to advance the learning of teachers both in the classroom and in collaborative groups. She is a former classroom teacher, elementary math specialist, and assessment coordinator. Katie strives to support teachers and teacher-leaders in their efforts to improve mathematics learning opportunities for all students. She is active in state and national mathematics organizations and has served on the Arizona Math Leaders Board of Directors. She received her bachelor's degree from Boston College and her master's degree from Northern Arizona University. Katie is a self-proclaimed "nerd" and proud of it; in addition to her love of numbers she is also an avid reader and enjoys walks on the beach with Honey, the family's Silver Labrador.

Chryste Berda is an adventurer, traveler, teacher, coach, consultant, author, and mom of two amazing Gen Z guys. She finds joy in singing at the top of her lungs and in the art of teaching. Chryste is energized by sharing her passion for learning with her colleagues as the district math coordinator and as a Regional VP for the Arizona Association of Teachers of Mathematics. She is intensely curious about students' thinking and spends much of her time listening to students explain their ideas. Chryste is an Honors program graduate from Western Oregon University with a bachelor's degree in Elementary Education and Interdisciplinary Studies with a specialty in Mathematics. She holds two master's degrees from Arizona State University—one in Curriculum and Instruction, the other in Educational Leadership. She has taught mostly math to students in Grades K–12, their teachers, and leaders (in addition to some other very fun courses) since 1998. Chryste is a native Oregonian who for the past twenty-ish years has lived with her family in Arizona's Valley of the Sun. She and her husband are raising two Arizona natives and aspire to become "snowbirds" in the future.

Introduction

Hello, friend!

Spoiler alert—everyone can be good at math.

Read that again. **Everyone** can be good at math. This is true for each of the students in your classroom, and it is true for **you** as well.

Yes, daily math instruction may look different and feel different than it did when you were a student. Maybe you loved math as a child; maybe you struggled. Either way, you recognize that the daily math instruction you have been charged with teaching involves understanding, flexibility, and the ability to explain and visualize abstract thinking.

The teacher's role has changed. No longer are we disseminators of information, holders of "all the knowledge" that we bestow to the students in our charge. Instead, we are moderators of learning. We guide student thinking to deeper levels of understanding by providing manipulatives, crafting questions, and guiding student inquiry.

What a wonderful opportunity we have been given! As teachers we have the unique privilege to foster a child's love of mathematics, of numbers, and of learning! And again—**everyone** can be good at math.

What an immense responsibility this is. Relying on the manner in which you were taught and your feelings about math (positive or negative) will no longer serve the needs of the students you teach. You are looking for resources and support to increase your own sense of math efficacy. Your desire is to increase your "toolbox."

Congratulations . . . this is the book you've been looking for!

So, who are we? We are coaches, mentors, authors, and above all else, teachers. For a collective of eighty-five-plus years, we have embraced the magic of making math come alive for children and their teachers, and we are thrilled to share this book with you. We believe deep-down that math makes sense. We believe that everyone can be good at math.

Being good at math is about so much more than memorizing equations. Thank you for accepting the challenge to see the math you teach differently. Thank you for seeking clarity. Most importantly, thank you for wanting what is best for all of the children you serve.

We're in this together!

Kimberly, Katie, and Chryste

What This Book Is About

This book is intended to help **you** help your children. It was created to provide you with images and descriptions of the way our students learn and think. The images focus on students' understanding of how and why math works, and they are explained in plain and simple terms.

Why This Matters

The strategies and operations that pop into your adult mind when you think "math" are likely abstract. They may involve symbols and procedures that were "gifted" to you by a caring adult, one who showed you a series of steps that you were then asked to memorize and apply. What we know about math today includes focused attention on developing a deep and sustainable understanding of problems, as well as the perseverance to solve them. Furthermore, understanding and perseverance must be intentionally developed by building on and making connections within the learner's own understanding.

How Mathematics Learning Happens

Deep understanding in mathematics evolves naturally when students use physical (concrete) materials such as counters, linking cubes, or even cars pulled from a toy box or fruit from a bowl. In this "doing" stage, learners are modeling strategies and operations with physical manipulation.

Similarly, learners can also use visual (pictorial) representations to help them "see" the math. This is when they create visuals that either represent concrete objects such as pictures or sketches, or they construct more abstract representations such as number lines or number bonds to explain their thinking. At first, student-created drawings are littered with detail. For example, when making sense of a problem such as *There were 10 bunnies on the hill, 3 bunnies hopped away . . .* the learner may draw complete pictures of bunnies with tails, big feet, and floppy ears. With the right prompting and support, students in this stage can understand that a circle, even one drawn in a tens frame, can represent a bunny. An important part of this mode centers on students making connections between the physical and visual representations.

By making connections between the physical and visual representations, students provide meaning that can also be connected to symbolic understanding, sometimes referred to as the "abstract mode." Here, learners use numbers and other mathematical symbols to model their thinking. A studio artist we know well once said when talking about human anatomy, "People think that they can make abstract art—they think it's easy to just throw paint and shapes on a canvas. A real artist, however, must understand the foundation: the location of the bones, the shape of the muscles. Only when you know how the form looks at its foundation can you make an abstraction of something such as the human form."

In this abstract form, learners can think critically about how they might represent math and their own understanding by basing it on and making connections between the different representations.

Note that as students make connections between and among physical, visual, and symbolic representations, their understanding deepens even further when they talk about their thinking (verbal representations) and connect their thinking to real life (contextual representations). It's the connections between each representation that lead to deeper understanding.

Figure i.1 shows an example of how students might connect representations for place value using base-ten blocks, sketches, and symbols.

Figure i.1 An Example of Concrete, Representational, and Abstract Representations of Place Value

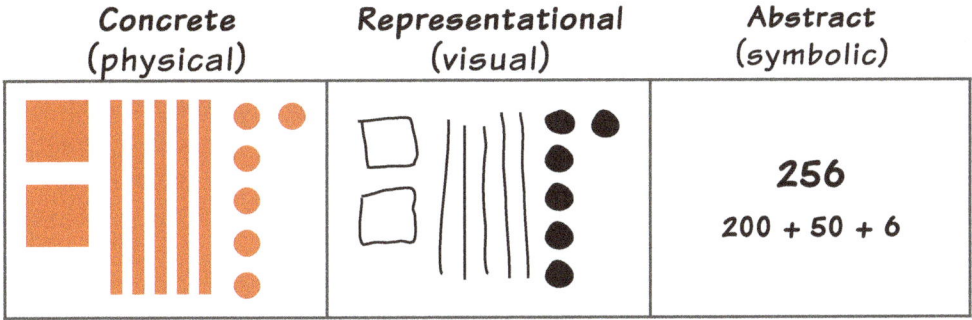

Figure i.2 shows an example for how learners might connect representations for equivalent fractions using two-color counters, sketches, and symbols.

Figure i.2 Concrete, Representational, and Abstract Representations of Equivalent Fractions

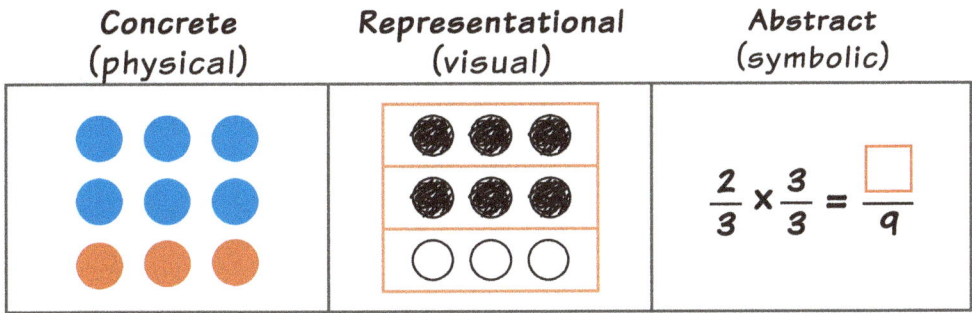

Figure i.3 show how students might connect representations for adding integers using algebra tiles, sketches, and symbols.

Figure i.3 Concrete, Representational, and Abstract Representations of Adding Integers

Introduction

Unfortunately, our school system is often steeped in doing things as they've always been done. This is especially true in math class. While there has been great effort in the last fifteen to twenty years to focus on improving conceptual understanding, there are still many classrooms in which students (and sometimes teachers) often rush to abstract and procedural mathematics, or "just numbers," without spending the time needed to develop deeper understanding with objects, tools, drawings, and the like. When this occurs, math learning is short-circuited as we resort to memorization rather than deep understanding.

Part of your role as a math teacher centers on developing your own understanding of the many ways both to see and represent the math you teach. By knowing and understanding these strategies, you are able to support and encourage students' thinking as they select and create their **own** representations. Only then do students make sense of mathematics; it is the **students** who should be choosing and creating the physical and visual representations, not simply mimicking what the teacher shows them. Please do your best to avoid "show-and-tell teaching"; instead, support students in the ways **they** choose how to show their thinking.

What the Role of Teacher Looks Like

A skillful teacher is able to probe for understanding by going back and forth between the physical, visual, and symbolic representations, using conversation to help the learner verbally craft connections between their thinking and the way they choose to represent it, and then connect those understandings to contexts in their own lives.

If you are most familiar and accustomed to one or maybe two particular algorithms or strategies for any given operation, this is the perfect opportunity for you to make your own connections. In the pages that follow you will find not only drawings and images of student thinking, but links to videos that model the use of manipulatives. We urge you to ask yourself, "How does this fit with what I know?" "How does this strategy mirror the strategies that I am familiar with using?" "What new ideas does this give me?" We are here to help you answer all of these questions.

Why These Representations?

It is important to note that we have chosen to model many of the most common strategies developed by learners, those that we see most often in the classroom. Kids, however, are always full of surprises. Some will be likely to invent their own strategies, which may be given equal consideration. Exploration and discussion of student strategies should not be seen as an opportunity for the teacher to **show** students what to "do" and ask that they replicate it. When we acknowledge student thinking and sense-making as a whole group by sharing strategies and asking thoughtful, probing questions, we empower all students.

And that is our most basic intent. We wish to empower you with additional knowledge and tools to enhance your own clarity and understanding in mathematics. We believe you will find what you're looking for here.

How to Use This Book

This is not the kind of book you read cover-to-cover. You will most likely use this book as a reference, jumping around as needed to inform the math teaching and learning going on in your classroom. We anticipated you would want to jump around, and we organized the book accordingly.

Let's take a minute to show you how this guide is put together.

There are sixteen color-coded chapters that can be used to quickly locate a topic, and you can jump around to find just what you need to know. Chapters begin with basic topics and go into more detail as you move toward the end. On pages ix–xiii, you'll see a list with all of the different math topics covered in this book. Look for the matching color on the edge of the page to go directly to the chapter.

There is also an additional tool at the back of the book to help you locate specific math problems: the **Topic Index**.

The **Topic Index** lists common math words alphabetically. This more typical way of listing common topics addressed throughout the book provides a handy way to look up topics such as "unit fractions" or "number lines."

Over the years, many teachers have found a variety of ways to use this book in supporting math teaching and learning. Some of those include using the book the following ways:

- As a preparation guide for self-learning about multiple ways to represent mathematical ideas.
- As a planning resource to anticipate a variety of student representations ahead of time.
- During PLC discussions with colleagues.
- To raise family awareness of how children may represent math when doing homework—often by copying a page or two and sending it home ahead of time.
- To support student awareness of representations they may choose from—again, often by copying (or projecting) a page or two and sharing them with students.
- As a reference to the math taught in your curriculum resources or textbooks to provide additional support for representing math thinking, especially in places where the textbook is heavy with "numbers only" work that would benefit from more student visuals.

These are just a few samples of how this book can be used. We encourage you to make it your own, finding new and innovative ways to support your own learning as well as that of your students and their families.

You will note that there are also several videos sprinkled throughout the volume. Since this book primarily centers on visual representations, we wanted to provide support for connecting visuals to physical objects (e.g., manipulatives). Therefore, these videos provide brief snippets for how you might help students connect physical, visual, and symbolic representations. As mentioned before, it's important that you use these to guide your thinking as you design instruction that facilitates students' thinking rather than resorting to "show-and-tell teaching."

We hope you find this to be a valuable resource to validate, reinforce, and expand your thinking and understanding . . . truly being able to "see" the math you teach so that your students can see it too.

Chapter 1

Whole Number and Decimal Place Value

Students come to school with a natural sense of quantity. They are born hardwired to compare quantities and determine which has more or which has less. That said, most students typically require direct opportunities with increasingly complex ideas about number and quantity to gain a greater understanding of our base-ten number system.

During the primary grades, the foundation for the base-ten system is built by developing a focused understanding of place value. This includes understanding that the entire base-ten system is made up of only ten digits (0–9). Once we have counted 0–9 to reach ten, we have a new unit made from a group of ten ones, called "a ten" for short. We represent this new unit by writing two digits, 1 and 0, side-by-side. This first grouping of ten may seem simple, but it's quite a complex idea for a young child.

As numbers increase in size and value, we record digits side-by-side to represent the number of groups in each "place." The ones place shows 0–9 units, the tens place shows how many groups of 10, the hundreds place shows how many groups of 100, and so on. Each group of three digits (ones, tens, and hundreds) is called a "period," and this pattern is repeated over and over to help us understand and read numbers. Each period includes a three-digit number, separated from other periods with a comma, to represent thousands, millions, billions, etc., that efficiently represents larger numbers.

Students develop an understanding that as you move left from digit to digit, the "magnitude" of each place is always ten times greater than the place to its right; and similarly, the magnitude of the place to the right is always one-tenth the value to its left. This sets them up to understand powers of ten relationships and decimals less than one. Without these understandings, students may struggle to develop more sophisticated ways of counting and comparing quantities as well as work more complex math operations.

Typical Trajectory in Most State Standards Frameworks:

- K–1: Whole numbers within 20
- Grades 1–2: Whole numbers within 100
- Grades 2–3: Whole numbers within 1,000
- Grades 4–5: Whole numbers within 1,000,000 and decimals through the hundredths places
- Grades 5–6: All whole numbers and decimals through the thousandths place

Ones and Tens on a Ten Frame

A **ten frame** is an array (group) of squares that can be easily used to visualize numbers between 0 and 10, which helps with visualizing how numbers can be combined or broken apart.

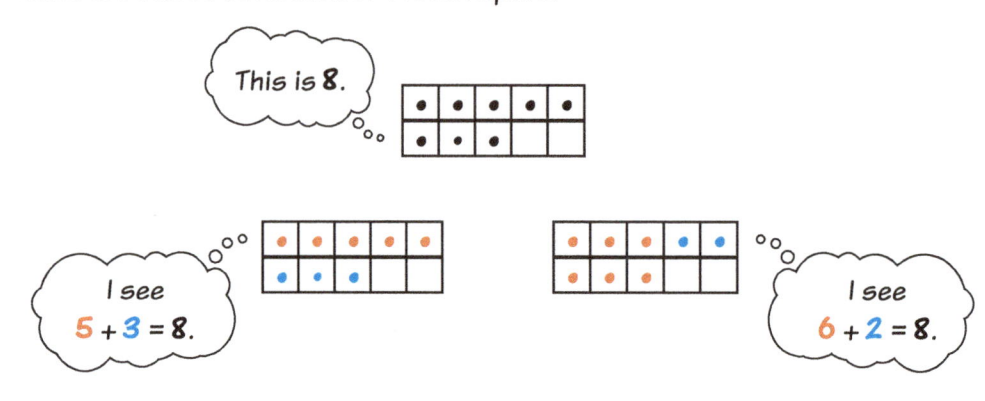

Unitizing is understanding that the number "ten" can be thought of as both 1 group of ten and 10 individual ones.

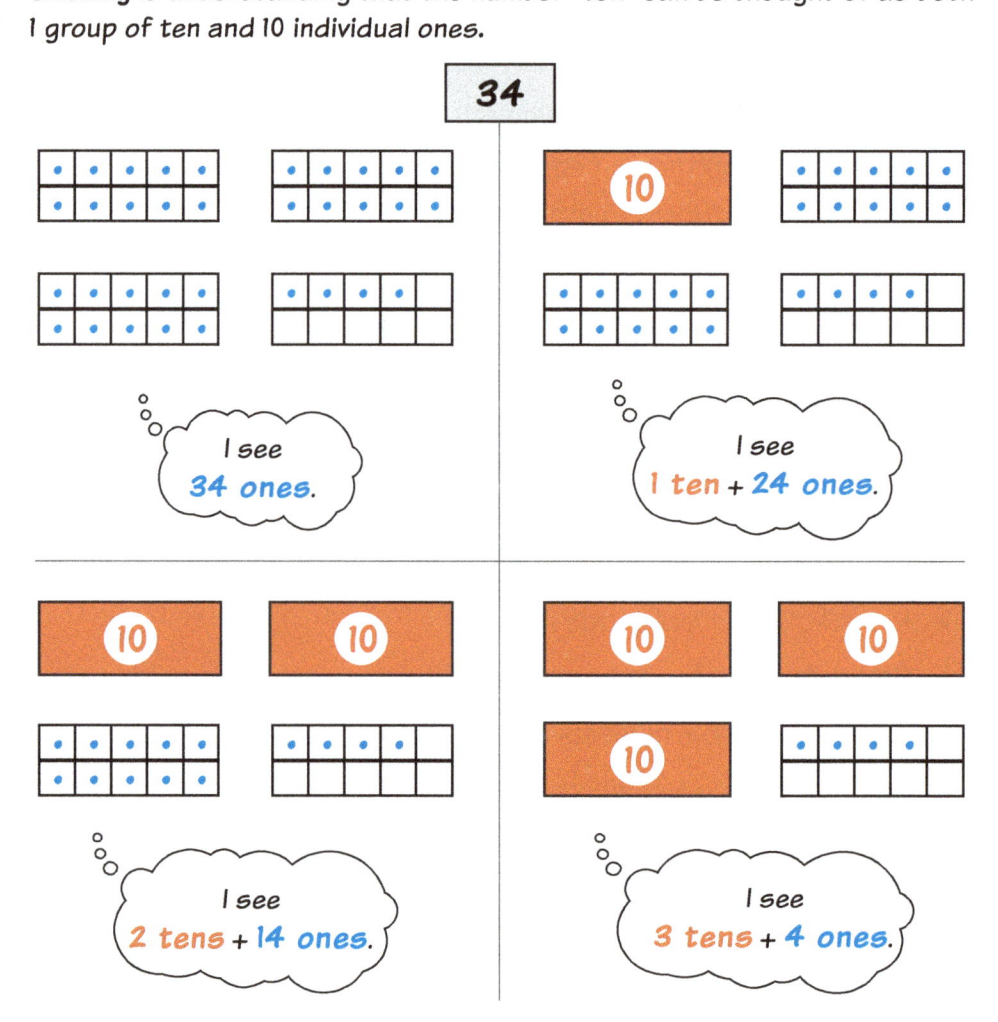

Whole Number Place Value

Multi-digit numbers can be understood by looking closely at the **place value** of each digit.

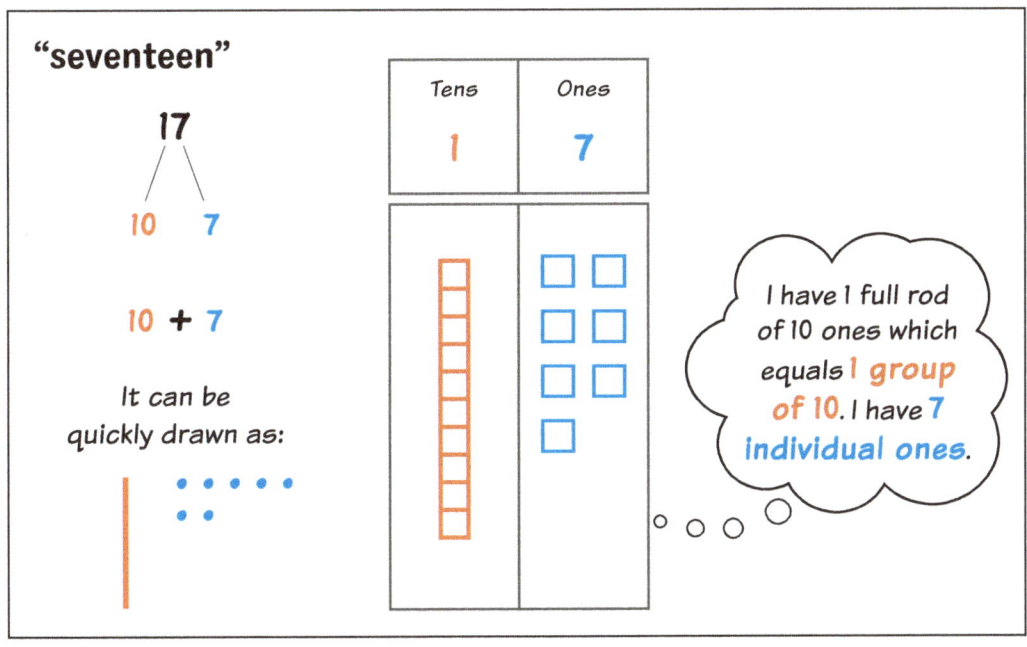

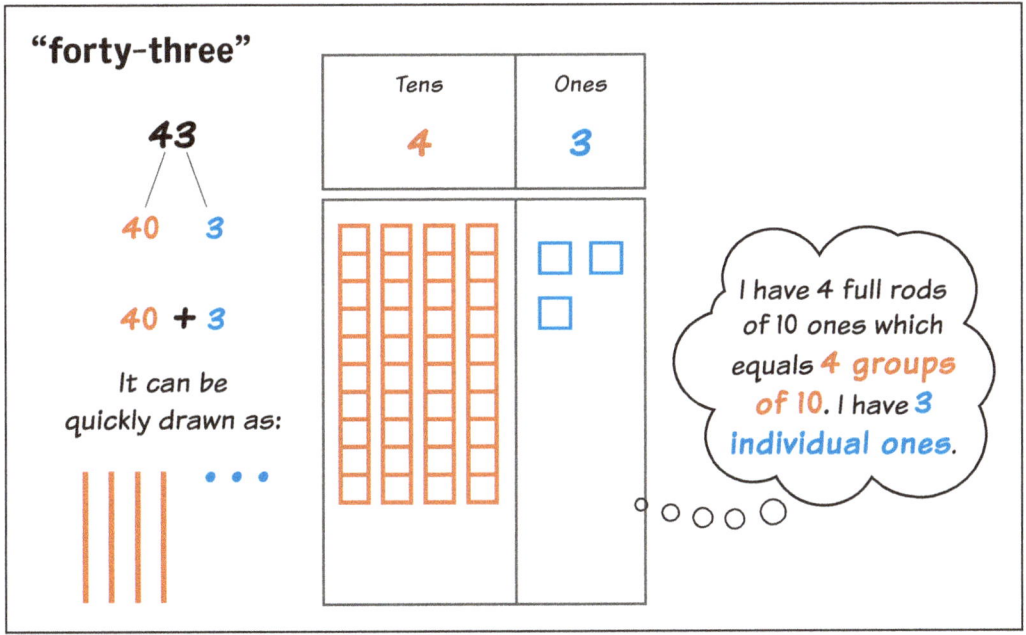

Whole Number and Decimal Place Value

Whole Number Place Value (continued)

Video 1: Place Value Pieces: You can use base-ten blocks, PV chips, and a variety of other place-value pieces to represent whole numbers, decimal fractions, and operations. *Pictured here: base-ten blocks, place-value disks, and KP® Ten-Frame Tiles*

https://qrs.ly/r3g99ke

To read a QR code, you must have a smartphone or tablet with a camera. We recommend that you download a QR code reader app that is made specifically for your phone or tablet brand.

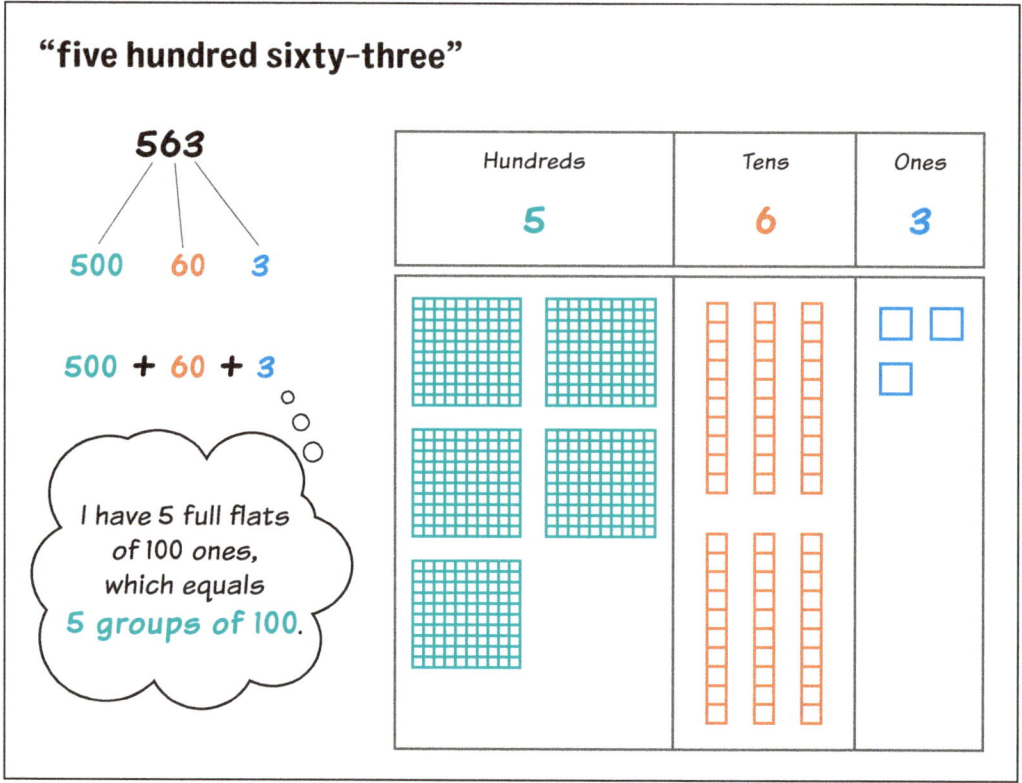

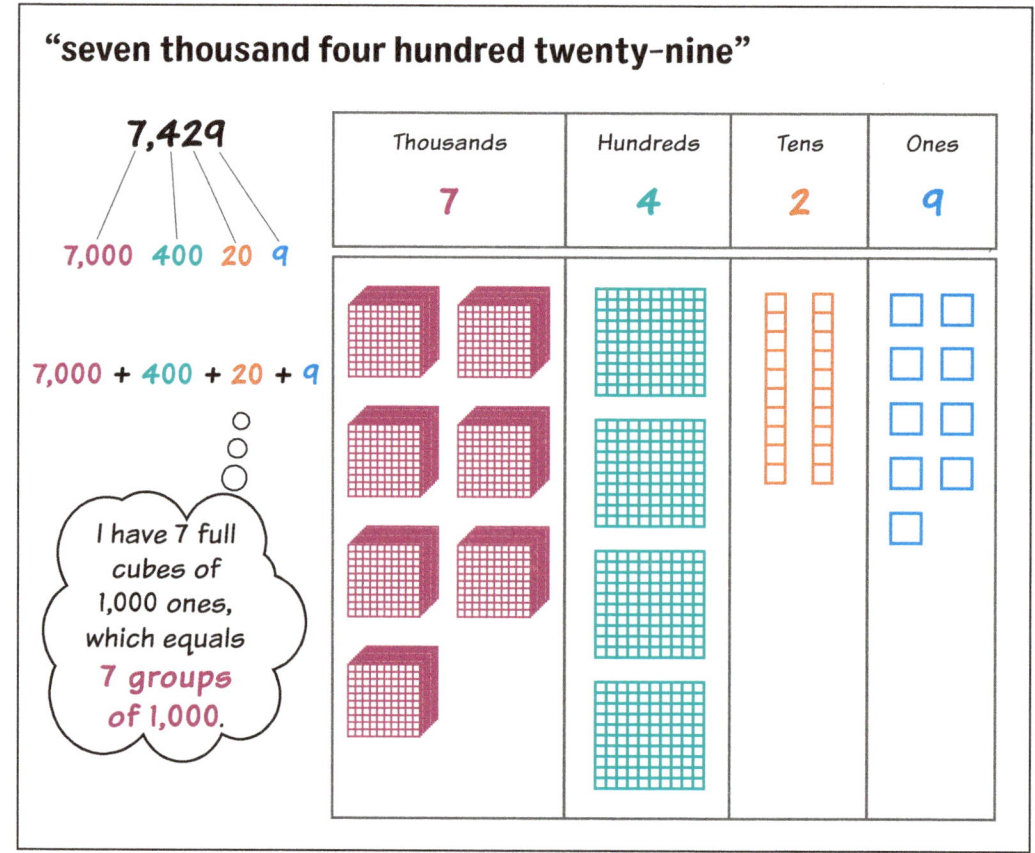

Whole Number Place Value (continued)

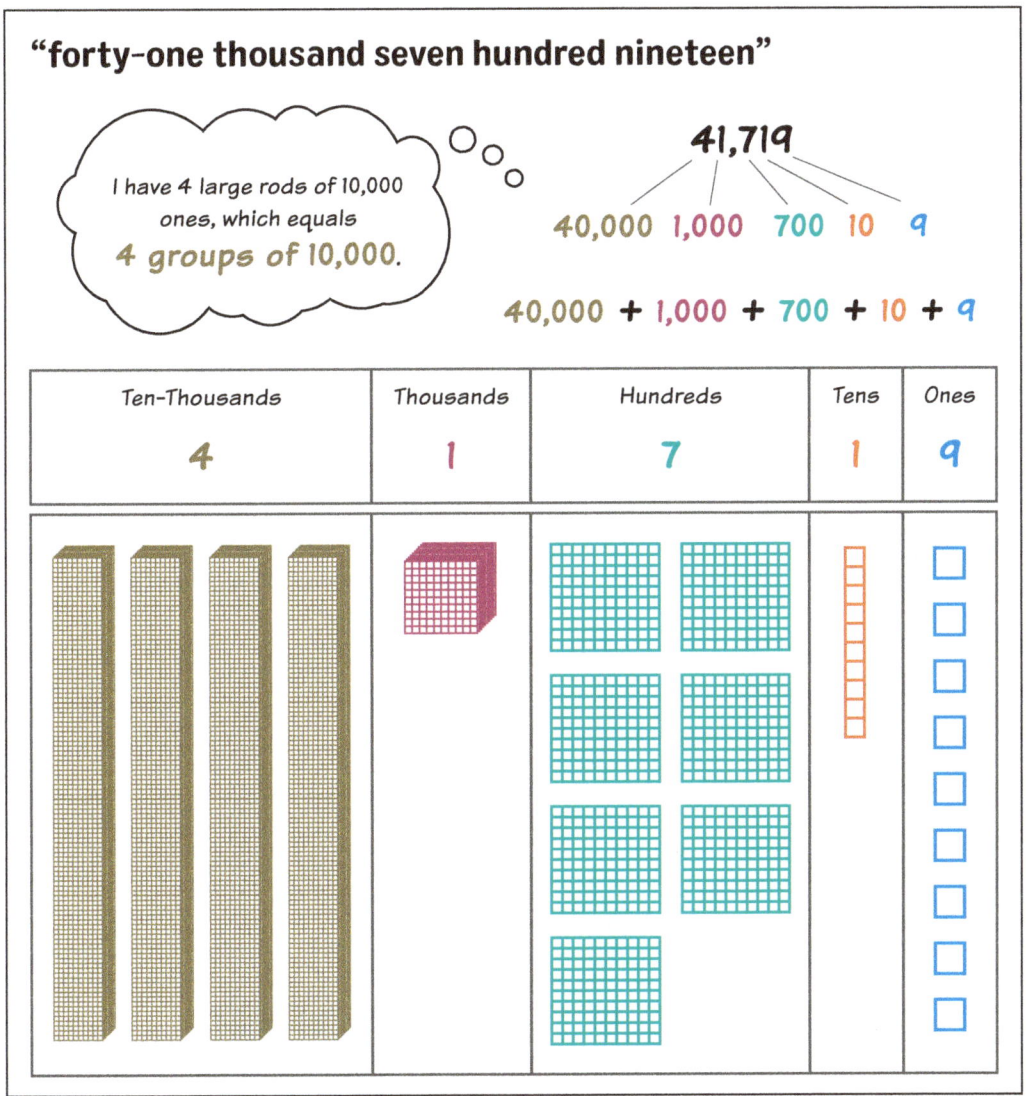

"forty-one thousand seven hundred nineteen"

I have 4 large rods of 10,000 ones, which equals **4 groups of 10,000**.

41,719
40,000 1,000 700 10 9

40,000 + 1,000 + 700 + 10 + 9

Ten-Thousands	Thousands	Hundreds	Tens	Ones
4	1	7	1	9

____ , ____ ____ ____ , ____ ____ ____

Millions — Hundred-thousands — Ten-thousands — Thousands — Hundreds — Tens — Ones

Place Value Organization

The base-ten system is made up of only ten digits (0–9). Once we have counted 0–9 and reached ten, we have a new unit made from a group of ten ones, called "a ten" for short.

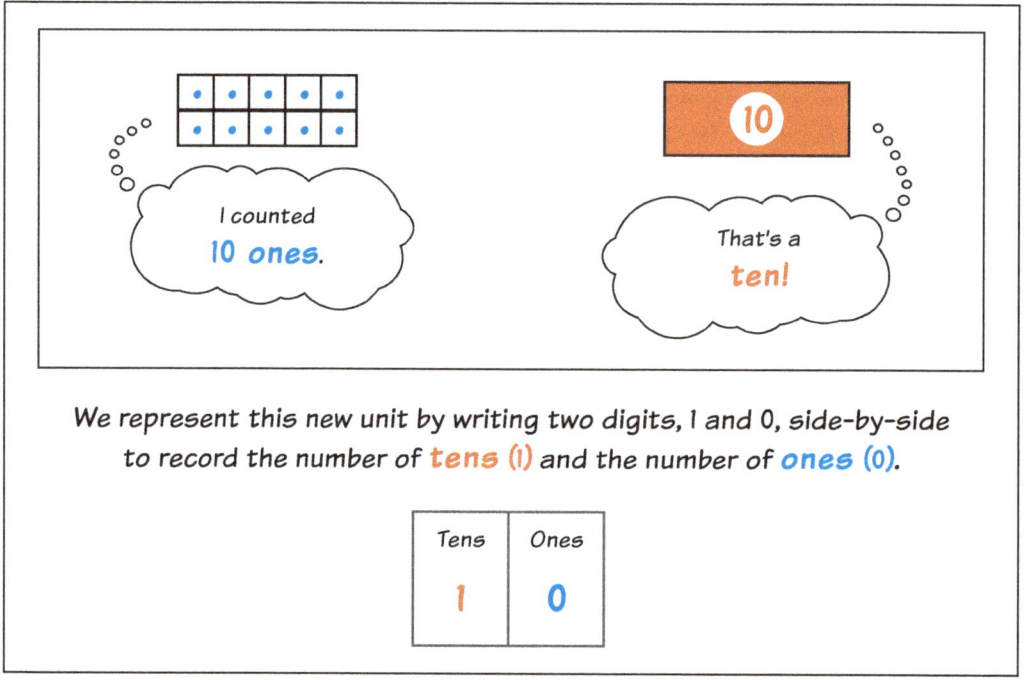

We represent this new unit by writing two digits, 1 and 0, side-by-side to record the number of **tens** (1) and the number of **ones** (0).

Tens	Ones
1	0

 What happens in the ones place, happens in every place. So when we've counted 9 tens and reached 10 tens, we have a new unit made from a group of 10 tens, called "a hundred" for short.

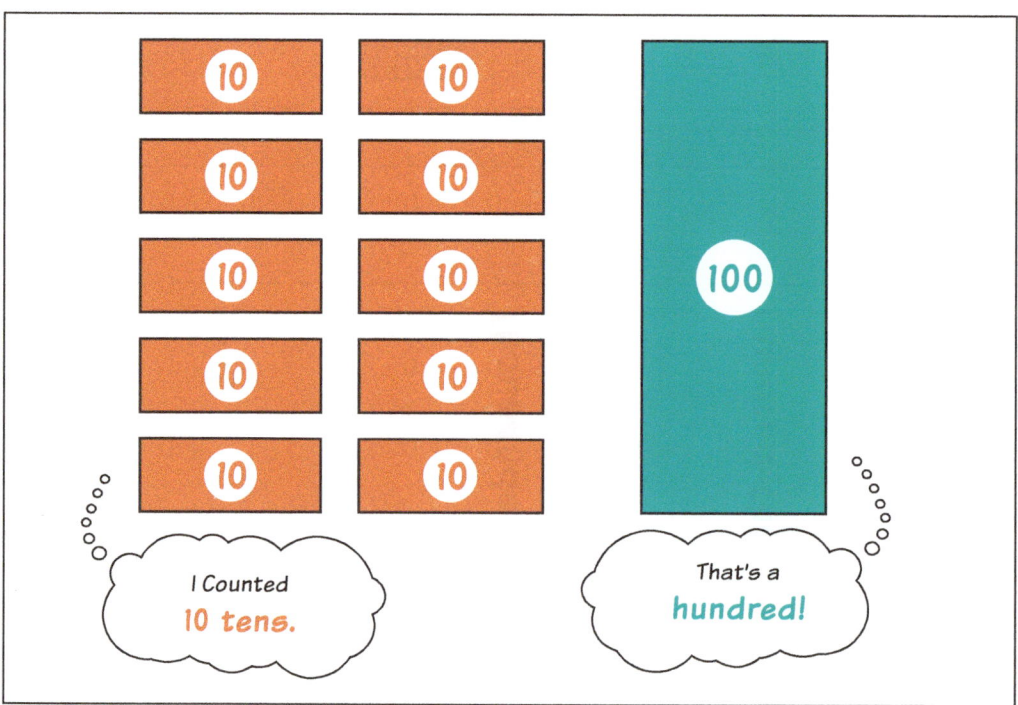

Place Value Organization (continued) ....

We represent this new unit by writing three digits, 1, 0, and 0, side-by-side to record the number of hundreds (1), tens (0), and ones (0).

Hundreds	Tens	Ones
1	0	0

The same is true for the periods. What happens in the ones period happens in every period. A **period** is a set of three digits representing groups of ones, tens, and hundreds of units that is repeated over and over with different magnitudes.

Millions Period				Thousands Period				Ones Period		
Hundreds	Tens	Ones		Hundreds	Tens	Ones		Hundreds	Tens	Ones
		1	,	0	4	1	,	7	1	9

Whole Number and Decimal Place Value

Decimal Place Value

You can use a **decimal** to represent a fractional part of a number that is less than one whole. Decimals are commonly used to represent money.

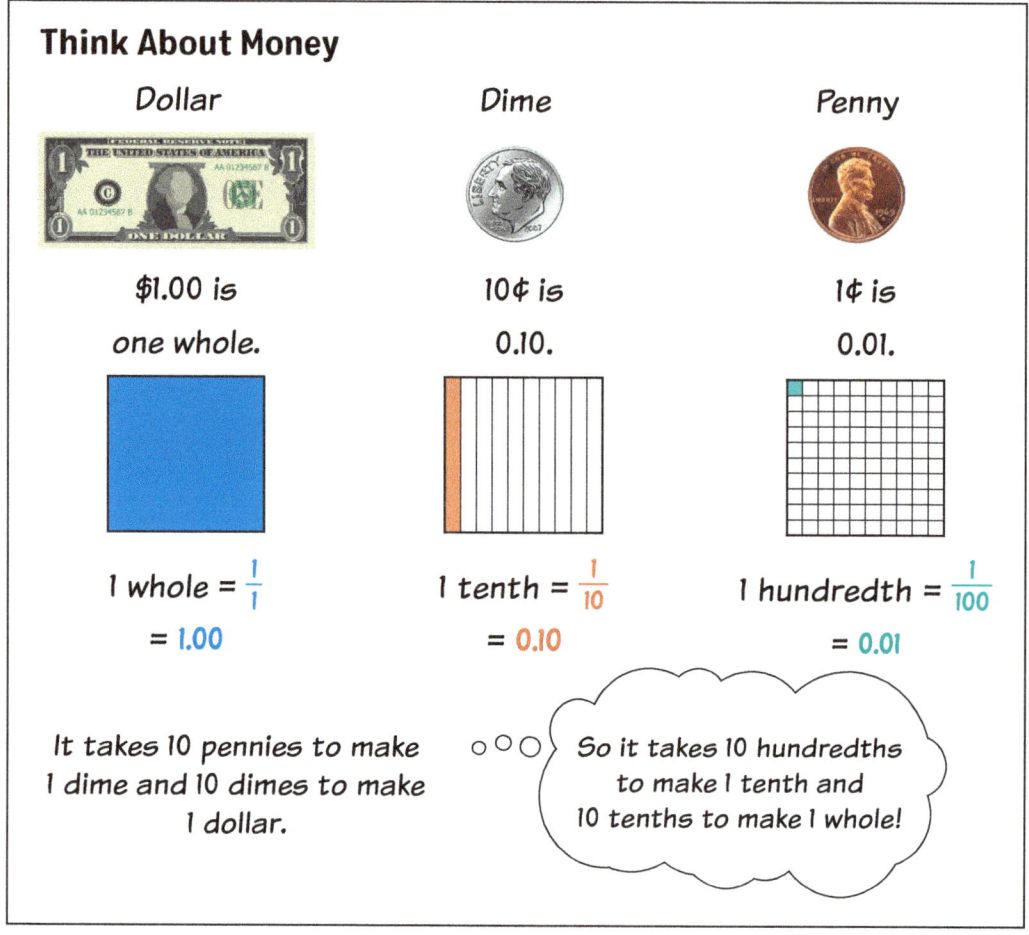

Source: Bill image by istock.com/SereiKorolko; Coin images by istock.com/filo

 The 10-to-1 relationship described with money works for all digits that are side-by-side in decimal place value as you move from right to left.

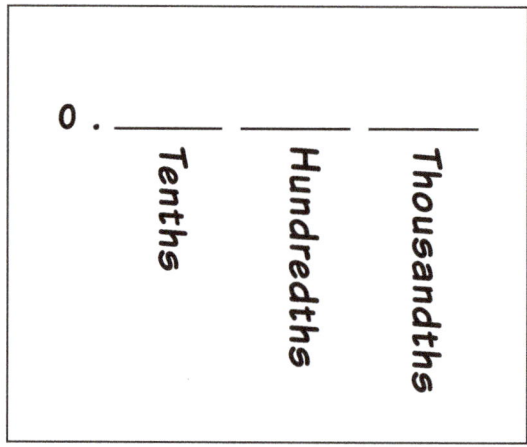

Decimal Place Value (continued)

A decimal number can be understood by looking closely at the **place value** of each digit and seeing that it takes ten of the value to the right to make one of the value immediately to the left. (You can use a **decimal** to represent the fractional part of a number that is less than one whole.)

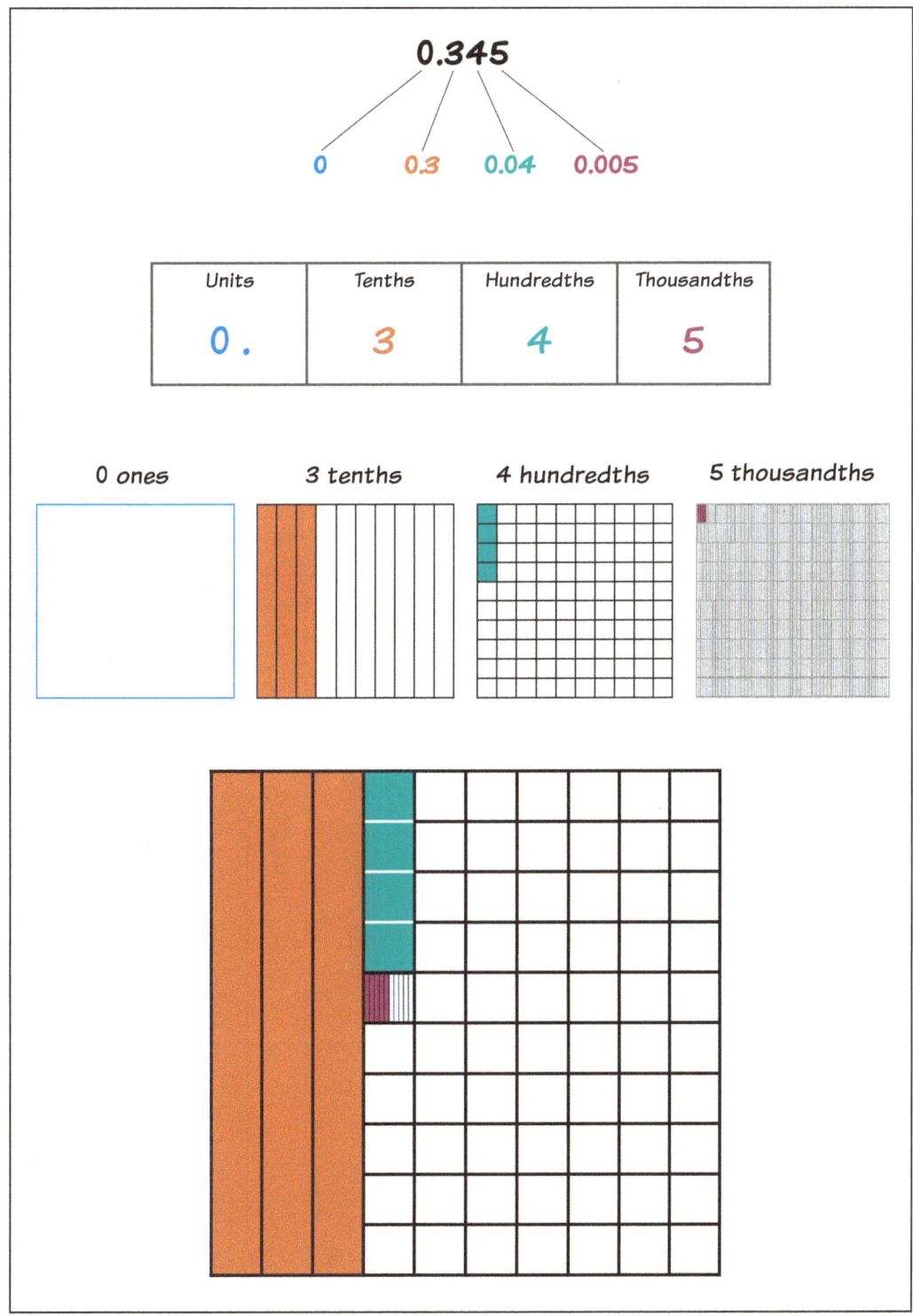

Decimal Point

Just as there is space between the whole number part and the fractional part of a number like $3\frac{12}{100}$, we use a **decimal point** to create space between the whole number and the fractional part of a number.

$3\frac{12}{100} = 3.12$

Whole Number Part		Fraction Part
3		$\frac{12}{100}$
3	.	12

I can read both of these numbers as "3 and 12 hundredths." So I know that there are 3 wholes and 12 hundredths of a whole.

💡 As digits are added to the left of the decimal point, we are representing counting more "things." As digits are added to the right of the decimal point, we are no longer counting more "things"; instead, the digits represent a single fraction that is less than one, and we are simply increasing the precision of that fraction.

Whole Number Part									Decimal Point	Fraction Part less than 1		
Millions Period			Thousands Period			Ones Period				Decimal Period		
Hundreds	Tens	Ones	Hundreds	Tens	Ones	Hundreds	Tens	Ones		Tenths $\frac{1}{10}$	Hundredths $\frac{1}{100}$	Thousandths $\frac{1}{1000}$
		1,	0	4	1,	7	1	9	.	3	4	5

Comparing Numbers..............................

Both large and small numbers can be compared using place value. The symbols >, =, and < are used to depict the comparisons (see page 22 for more about math symbols).

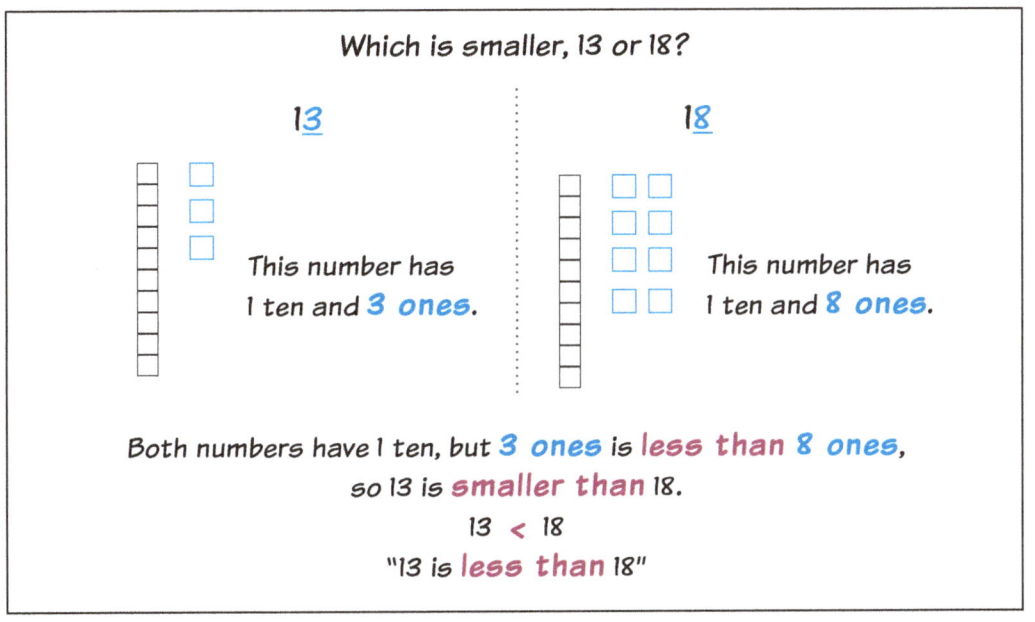

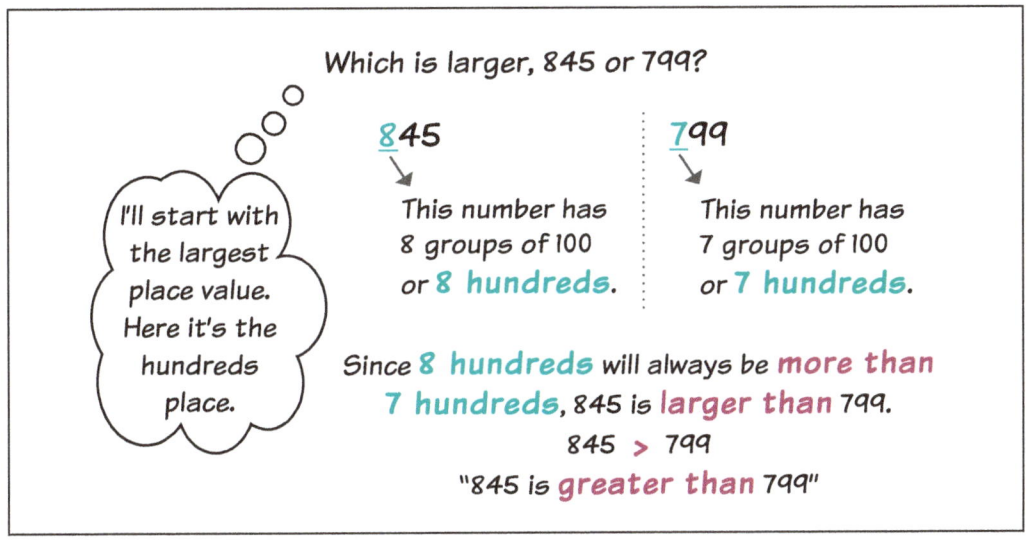

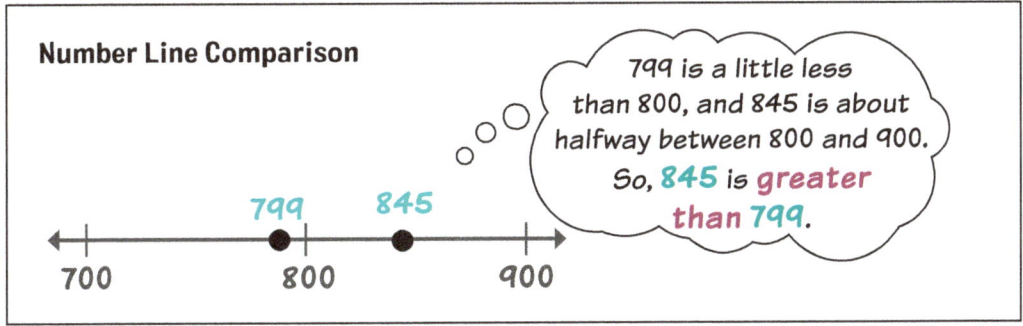

Whole Number and Decimal Place Value

Comparing Decimals

Because the number system is uniform, students can use the same place value understanding for comparing larger numbers and smaller numbers.

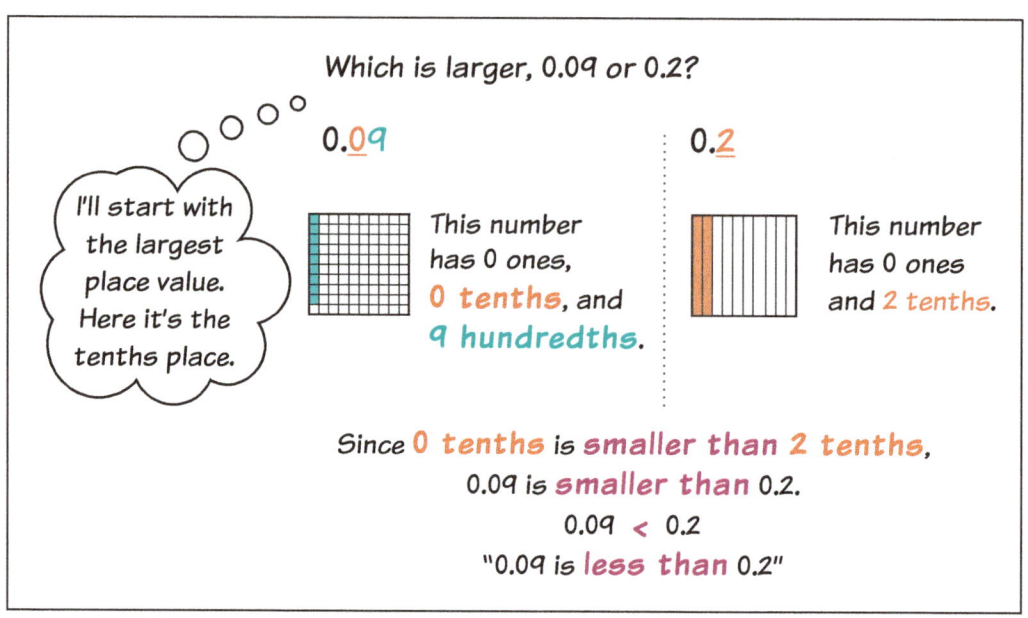

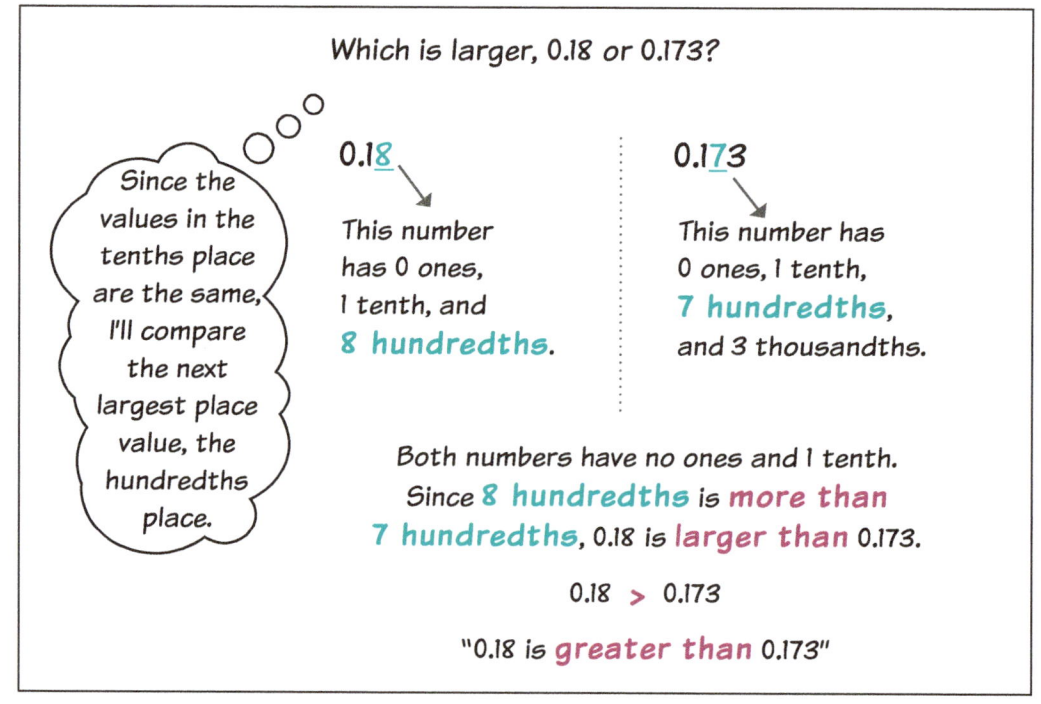

Chapter 2
Math Symbols and Properties

Just as words are used to communicate linguistically, math symbols are used to communicate mathematically. Each symbol conveys a meaning and should be used precisely to reflect the ideas being communicated.

Similar to verbs in language, math symbols indicate the actions or relationships being communicated. For example, whether comparing the value of quantities (=, <, or >), combining quantities (+), finding the difference between quantities (−), making copies of a quantity (×), or fairly sharing a quantity (÷), the math symbol succinctly communicates this to the mathematician.

Similarly, math properties are agreed-upon rules that help us communicate in a way that ensures we share our ideas about complex concepts clearly and can use these to solve problems.

Providing experiences where students communicate symbolically should stem from connections among the various representations of math concepts. For example, grounding symbols to real-world or contextual situations and diagrams or visual models helps students develop a more precise use of math symbols and properties.

Typical Trajectory in Most State Standards Frameworks:

- Grades K–6: Equality symbols (=, <, or >) for given number sets (see Chapters 1 and 7)
- Grades K–2: Addition and subtraction (+, −) for given number sets (see Chapters 4–9); commutative property, associative property
- Grades 3–6: Addition, subtraction, multiplication, and division (+, −, ×, ÷); commutative property, associative property, distributive property, identity property

Commonly Used Math Symbols..............

Math symbols are used to communicate in math, just as words are used to communicate in language. Each symbol conveys a special meaning and should be used precisely.

Symbol	Read As	Explanation/Example	
=	is equal to, is the same value or quantity as	$3 + 5 = 8$ 3 + 5 is the same value as 8.	
≠	is NOT equal to, is NOT the same value or quantity as	$4 + 2 \neq 10$ 4 + 2 is NOT the same value as 10.	
+	plus, add, addition	$12\frac{1}{2} + 8 = 20\frac{1}{2}$ $12\frac{1}{2}$ plus 8 is the same value as $20\frac{1}{2}$.	
-	minus, take away, subtract	$9 - 4 = 5$ 9 minus 4 equals 5. 9 take away 4 is the same as 5 subtract 4 from 9 and the value is 5.	
×, ·, *, ( )	times, is multiplied by, groups of	$3.5 \times 7 = 24.5$ 3.5 groups of 7 is equal to 24.5.	$14(b) = 14b$ 14 times b is the same value as 14b.
÷, /, $\frac{x}{y}$, $\sqrt{}$	divided by, divided into groups	$6.8 \div 2 = 3.4$ 6.8 divided into 2 groups is the same quantity as 3.4.	$\frac{3}{4}$ 3 divided by 4
<	is less than	$19 < 20$ 19 is less than 20.	
>	is greater than	$14.25 > 14.193$ 14.25 is greater than 14.193.	
( )	the quantity of	$15 - (7 + 3) = 5$ 15 minus the quantity of 7 plus 3 is the same value as 15 minus 10, or 5.	

Signs of Comparison

The **equal sign** shows how the values or quantities on either side of the sign relate to each other. In addition to the word "equals," you can say "is the same value as" or "is the same quantity as" for the equal sign when you read an equation.

When comparing two quantities that are the same amount, you could put a board on top of stacks of the two quantities and the board would balance and stay straight because the two quantities **are the same** as each other. They are equal. The boards would look like an equal sign.

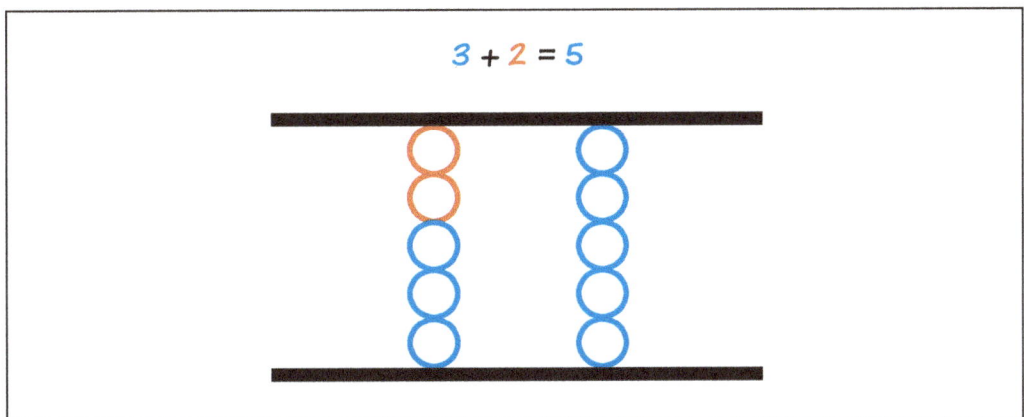

$3 + 2 = 5$
$5 = 5$

"Three plus two is the same quantity as five."

$6 = 7 - 1$
$6 = 6$

"Six is the same value as seven minus one."

$15 + 9 = 12 + 12$

"Fifteen plus nine is the same quantity as twelve plus twelve."

$4 + 1 = x + 3$

"Four plus one is the same value as some number plus three."

$15 + 9 = 12 + 12$
▲
"is the same as"

$4 + 1 = x + 3$
▲
"is the same as"

Signs of Comparison (continued)..............

$5 = 5$
$3 + 2 = 5$
$5 = 3 + 2$
$3 + 2 = 3 + 2$
$2 + 3 = 3 + 2 = 5$
$6 - 1 = 5$

These values and quantities are all equal because the values are the same on either side of the equal sign. It doesn't matter which side of the equal sign the expression appears on.

When comparing two quantities, you could put a board on top of stacks of the two quantities to know which is less or greater and which sign to use to record which is less or greater.

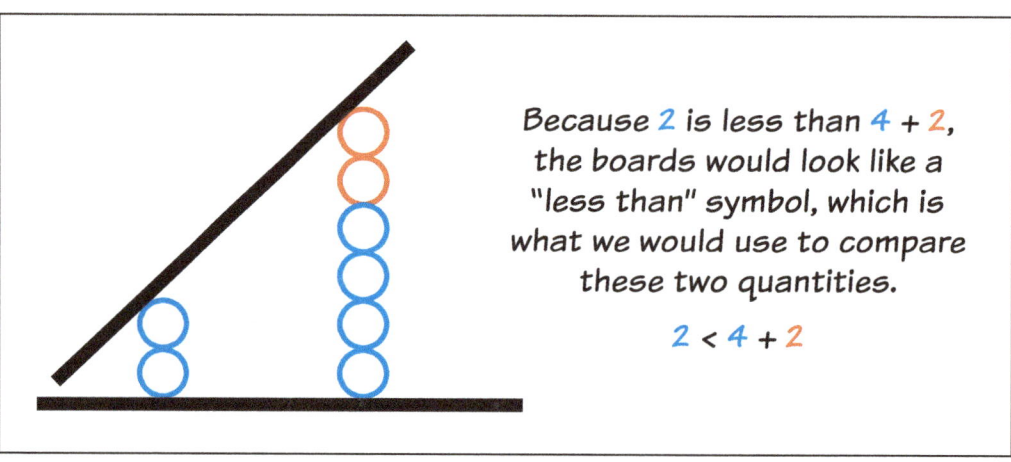

Because 2 is less than 4 + 2, the boards would look like a "less than" symbol, which is what we would use to compare these two quantities.

$2 < 4 + 2$

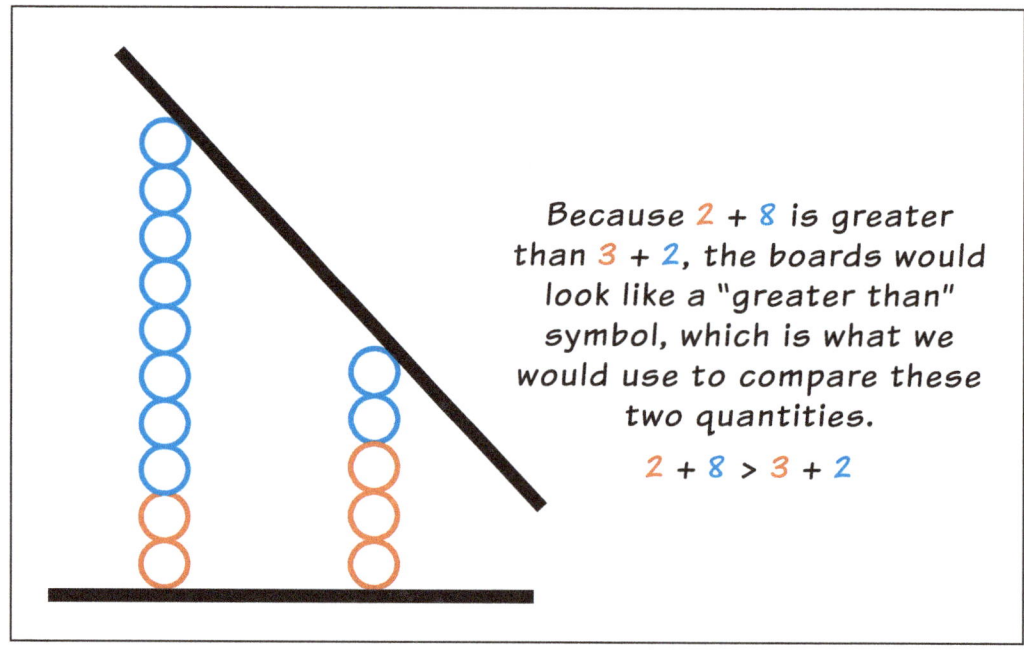

Because 2 + 8 is greater than 3 + 2, the boards would look like a "greater than" symbol, which is what we would use to compare these two quantities.

$2 + 8 > 3 + 2$

Properties of Addition

The **properties of addition** are math rules that make addition work. They are often discovered in primary grades and then used heavily in algebra.

Identity Property of Addition

$$a + 0 = a$$

When adding 0 to any number, the number does not change.

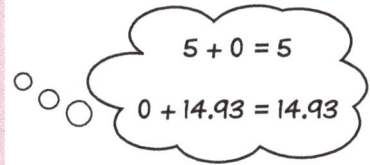

$5 + 0 = 5$

$0 + 14.93 = 14.93$

Commutative Property of Addition

$$a + b = b + a$$

When two numbers are added, the sum is the same regardless of the order of the addends.

$3 + 4 = 4 + 3$

$12\frac{1}{2} + 4\frac{3}{4} = 4\frac{3}{4} + 12\frac{1}{2}$

Associative Property of Addition

$$(a + b) + c = a + (b + c)$$

When three or more numbers are added, the sum is the same regardless of how the numbers are grouped.

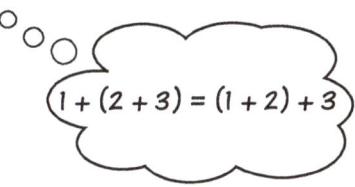

$1 + (2 + 3) = (1 + 2) + 3$

Remember that the = sign means "is the same as." So 5 + 0 is the same as 5.

OR

3 + 4 is the same as 4 + 3.

Math Symbols and Properties

Properties of Multiplication

The **properties of multiplication** are math rules that make multiplication work. They are often discovered in intermediate grades and then used heavily in algebra.

Identity Property of Multiplication

$$a \times 1 = a$$

When multiplying any number (or variable) by 1, the product is the number (or variable) itself.

Commutative Property of Multiplication

$$a \times b = b \times a$$

When two numbers are multiplied, the product is the same regardless of the order of the numbers.

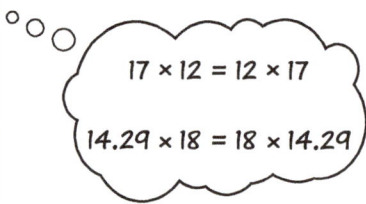

Associative Property of Multiplication

$$(a \times b)c = a(b \times c)$$

When three or more numbers are multiplied, the product is the same regardless of how the numbers are grouped.

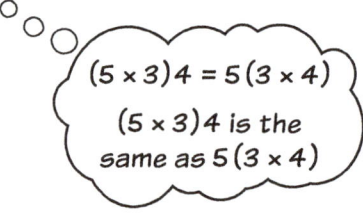

Distributive Property of Multiplication Over Addition

$$a(b + c) = (a \times b) + (a \times c)$$

When multiplying a sum, you can multiply each addend separately and then add the products.

Order of Operations

The **order of operations** ensures that we all follow the same steps when solving multi-step number problems.

The rules:

① First, do the operations in grouping symbols.

② Second, do the exponents.

③ Third, multiply and divide from left to right.

④ Fourth, add and subtract from left to right.

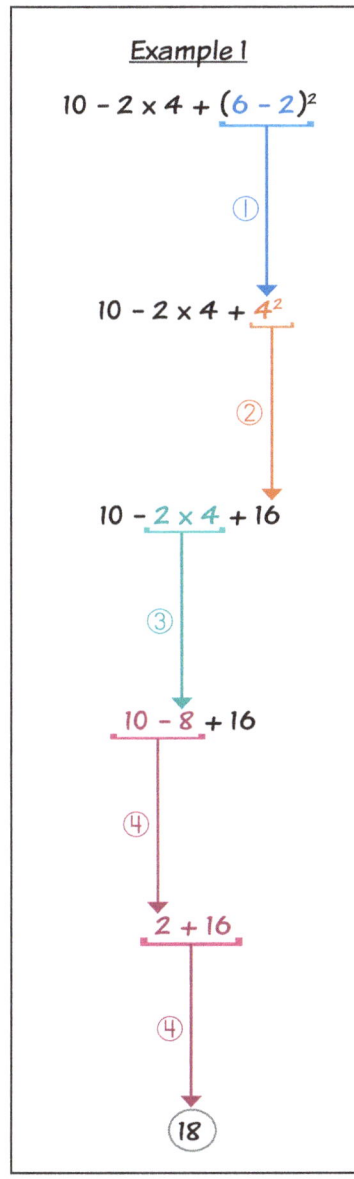

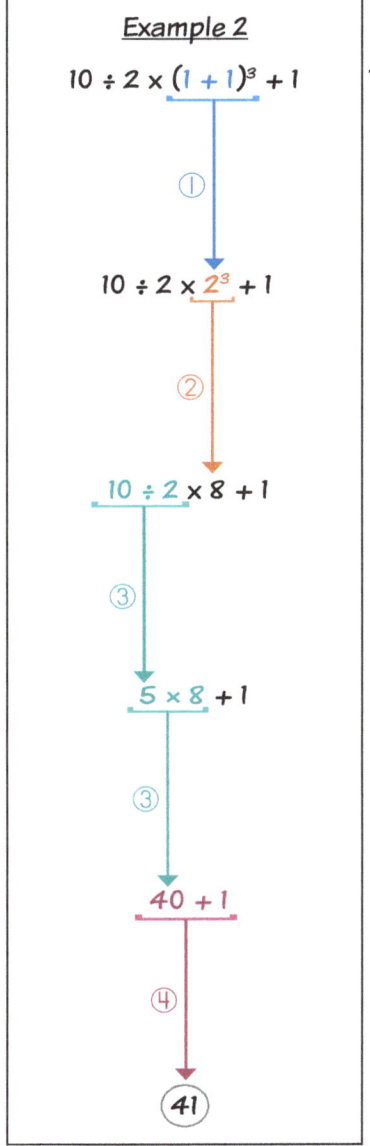

- Grouping Symbols
- Exponents
- Multiplication and Division (in order left to right)
- Addition and Subtraction (in order left to right)

Math Symbols and Properties

27

Chapter 3

Estimation (Including Rounding)

Estimation is not only a tool for math but also a tool for life. We use estimation when we don't need an exact answer—just an answer that is "close enough."

In life, we could use estimation in a lot of different situations, such as

- a close estimate for how much our groceries will cost,
- a quick way to know "about how much" change we should get back, or
- a fast guess about how much flour we'll need for baking four batches of cookies.

In math, we could use estimation to know "about how much" an answer to a problem should be, so that we can decide if the answer is reasonable. For example, when working with whole numbers, one can think that 51 – 29 is about 50 – 30, so the exact answer will be close to 20. Or, when operating with decimal fractions, students can use estimation to realize that 3.0125 + 1.9 is about the same value as 3 + 2, so the exact answer will be close to 5.

Estimates help students make informed decisions, using mental math strategies, about how to solve problems efficiently. They also use estimation to make generalizations and assess reasonableness of visual displays like charts and graphs and reasonableness of computations.

Note that although "rounding" gets a lot of attention in our classrooms, it is only one of many methods for estimating. Other approaches include front-end estimation, chunking, and using benchmark numbers.

Typical Trajectory in Most State Standards Frameworks:

- Grades K–6: Students use appropriate estimation strategies throughout the elementary grades (rounding typically gets attention in Grades 3–4, but it's often taught algorithmically rather than as a way to find a solution that is close)

Estimation (Including Rounding)

Video 2: Rounding With Virtual Number Line and Ten Frames: You can use virtual number lines and manipulatives to help students grasp the idea of rounding as "closer to" rather than teaching rounding rules through songs and rhymes. *Virtual number line and ten frames from the Math Learning Center*

https://qrs.ly/fwg99kb

Link to Math Learning Center Apps:

https://qrs.ly/7bg99jn

To **estimate** is to find a number close to an exact amount. This is done by determining what **benchmark** or **friendly number** it is closest to. A **benchmark number** is a number that ends in either 0 or 5. A **friendly number** is a number that is easy to work with.

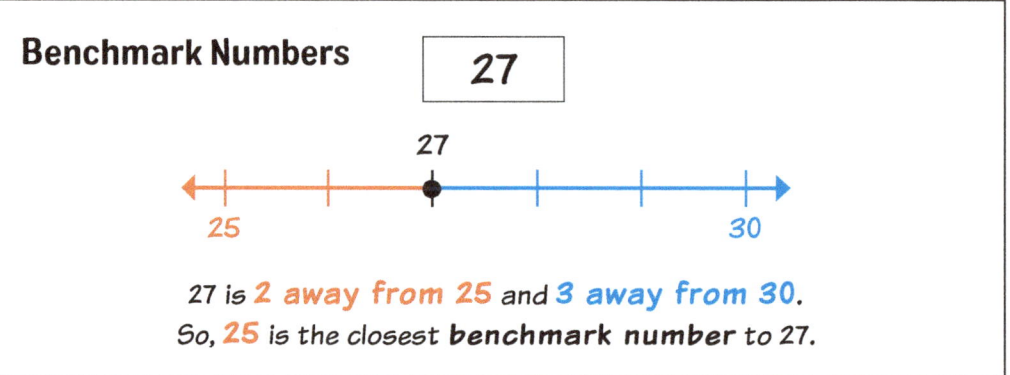

Benchmark Numbers

27 is **2 away from 25** and **3 away from 30**.
So, **25** is the closest **benchmark number** to 27.

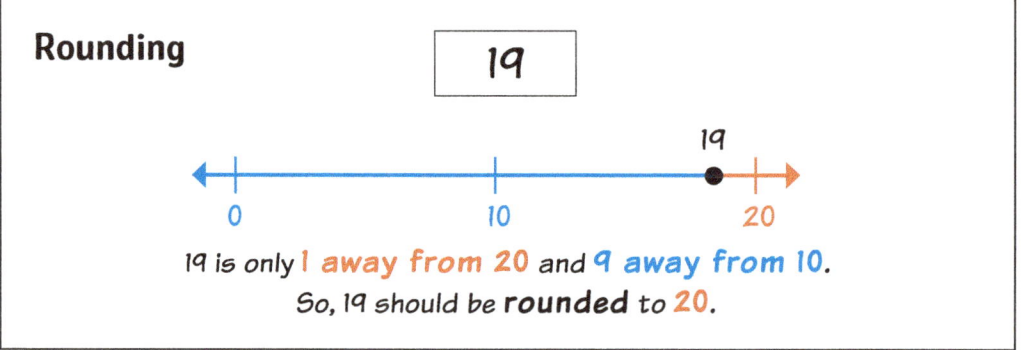

Rounding

19 is only **1 away from 20** and **9 away from 10**.
So, 19 should be **rounded** to **20**.

 Estimating helps us decide if our answers are reasonable.

*I know that 12 is close to 10 and 19 is close to 20. I could **estimate** the answer to be about 10 + 20, which is 30.*

Actual: 12 + 19 = 31

Estimate: 10 + 20 = 30

An open number line starts as a line with no numbers. Students are able to mark points or numbers to record their thinking during the process of mental computation. An open number line extends in both directions with an arrow to indicate that numbers never end. They go on infinitely in both directions! Students are able to use an open number line to visualize how "close" any given number is to more familiar benchmark numbers.

Rounding Large Numbers

Numbers can be rounded to different place values. Sometimes it is necessary to round a number to a specific **place value**.

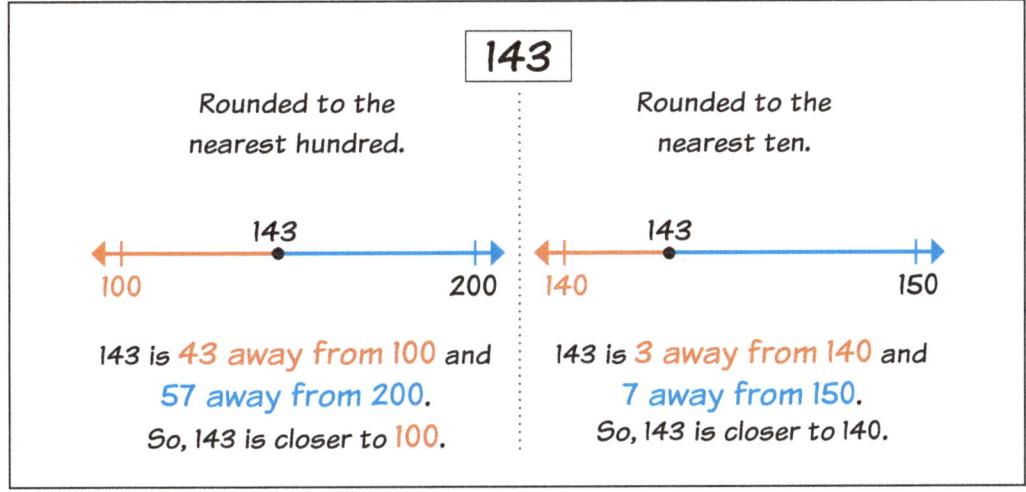

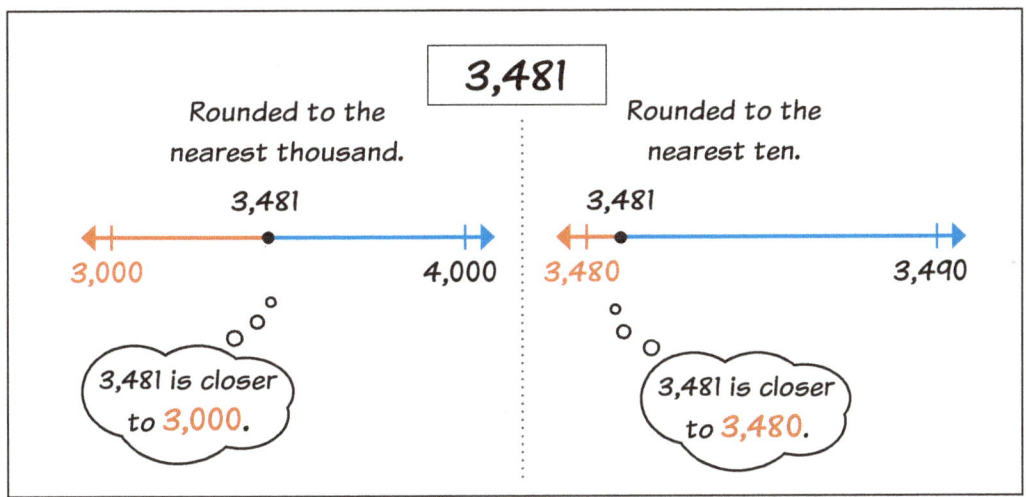

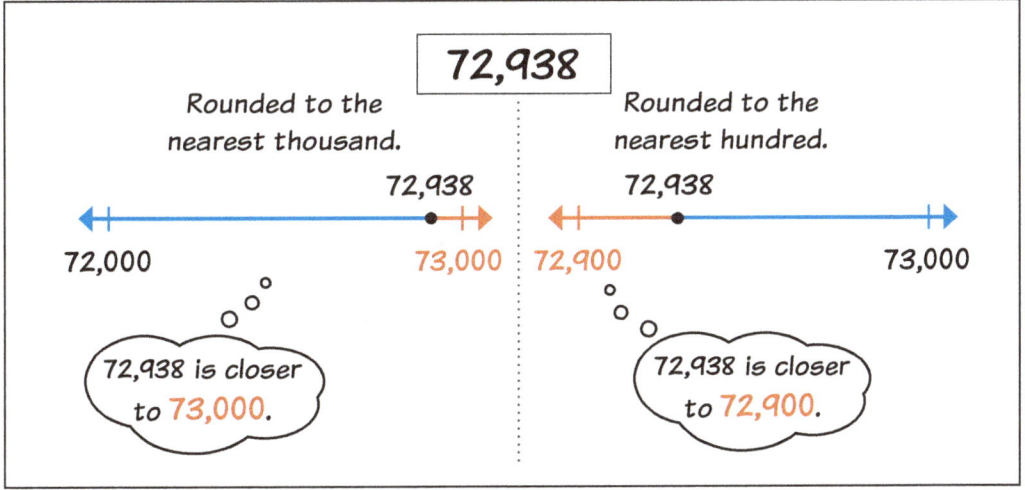

Rounding With Decimals

Decimal numbers can be rounded to whole numbers or to a specific *decimal place value*.

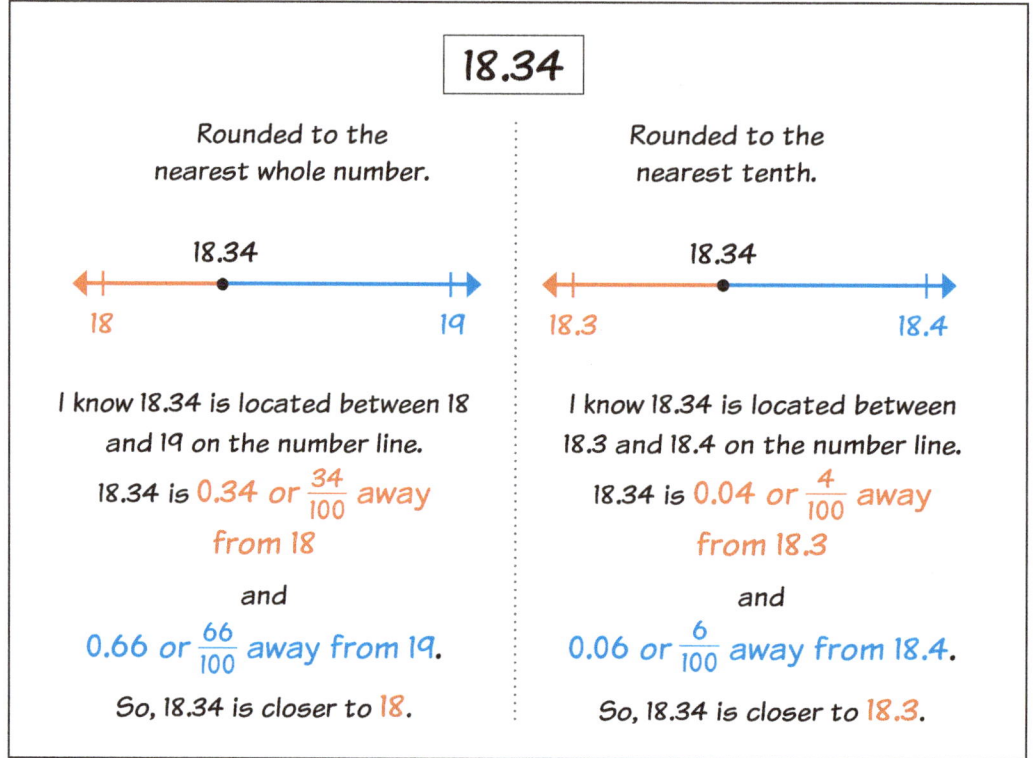

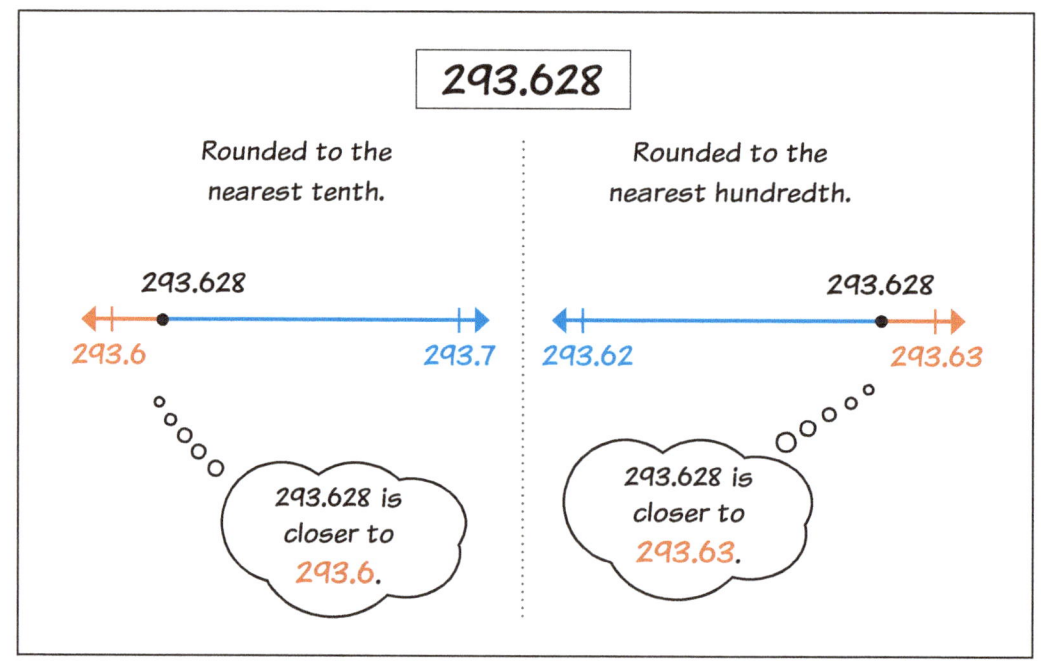

Estimation Strategies

Although "rounding" gets a lot of attention in our classrooms, it is only one of many methods for estimating. Other approaches include front-end estimation, compensation, and using benchmark numbers.

Front-End Estimation

$$
\begin{array}{rcr}
\$ 3.95 & \rightarrow & \$ 3.00 \\
\$ 7.89 & \rightarrow & \$ 7.00 \\
+\, \$ 10.29 & \rightarrow & +\, \$ 10.00 \\
\hline
\$ 20.00 & & \$ 20.00
\end{array}
$$

I can keep the first digit the same and change all others to zeros.

Compensation Estimation

This estimation strategy is like front-end estimation, except that the remaining amounts are also considered and added to the estimate to get an estimate that is closer to the exact answer.

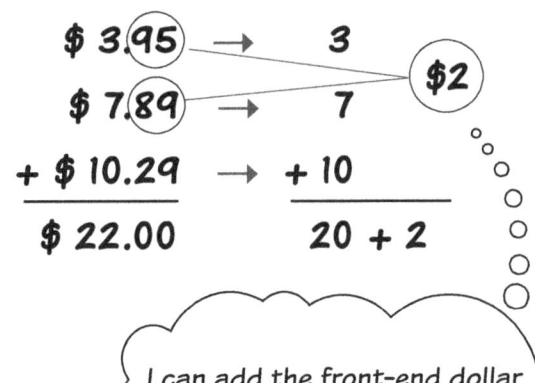

$$
\begin{array}{rcr}
\$ 3.\boxed{95} & \rightarrow & 3 \\
\$ 7.\boxed{89} & \rightarrow & 7 \\
+\, \$ 10.29 & \rightarrow & +\, 10 \\
\hline
\$ 22.00 & & 20 + 2
\end{array}
$$

I can add the front-end dollar amounts and then group the cents to form dollars.

Estimating With Benchmark Fractions

Like a benchmark number, a **benchmark fraction** is an easy fraction to work with, such as 0, $\frac{1}{4}$, $\frac{1}{2}$, $\frac{3}{4}$ and 1.

Estimating Using Logic

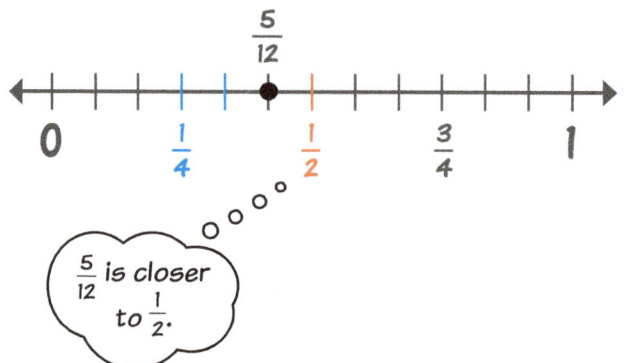

$\frac{5}{12}$ is closer to $\frac{1}{2}$.

Estimating Using Equivalent Fractions

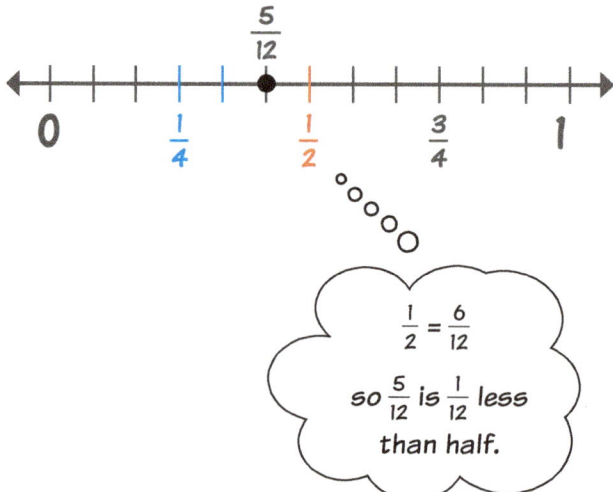

$\frac{1}{2} = \frac{6}{12}$ so $\frac{5}{12}$ is $\frac{1}{12}$ less than half.

Chapter 4

Addition and Subtraction Using Place Value

Addition and subtraction are often taught close together because they are inverse operations. This means that the operations are opposites and can "undo" each other. For example, addition is the "putting together" of two or more quantities, but adding can be undone by subtracting one of the quantities from the total to arrive at the starting amount.

When students understand this inverse relationship, they can think of subtraction problems as finding the unknown addend, interpreting a problem like $10 - 5 = ?$ as if it were $5 + ? = 10$. Because subtraction is as unnatural as addition is natural, students may find it easier to view subtraction as a missing addend question.

Similarly, in middle school, students can use this relationship to think of $10 - (-7) = ?$ as $-7 + ? = 10$, or interpret it as "beginning with 10 and doing the opposite of subtracting 7, which is adding 7, to find the difference between 10 and −7."

The first step to establishing the relationship between addition and subtraction is to begin with real-world situations, including word problems and other contexts. Using situations students can relate to helps them make sense of the inverse relationship—they can visualize the action of the problem in context.

When learning to add and subtract, students use a variety of strategies. Young students typically begin with counting strategies using physical objects and visual tools (e.g., pictures, sketches, virtual manipulatives, etc.) prior to and alongside symbols. Counting strategies are then replaced by reasoning strategies and eventually lead to fluency with math facts as well as with multi-digit numbers.

Typical Trajectory in Most State Standards Frameworks:

- Grade K: Adding and subtracting within 10 using conceptual strategies
- Grade 1: Adding and subtracting within 20 and within 100 using conceptual strategies
- Grade 2: Adding and subtracting within 20, within 100, and within 1,000 using conceptual strategies
- Grade 3: Adding and subtracting within 1,000 using conceptual strategies
- Grade 4: Adding and subtracting within 1,000,000 using a standard algorithm
- Grades 5–6: Adding and subtracting with all whole numbers using a standard algorithm and with decimals through the thousandths place using conceptual strategies

Video 3: Addition and Subtraction With Two-Color Counters: You can use two-color counters to help students "see" how the operations of addition and subtraction "behave" when using different interpretations for addition and subtraction. *Pictured here: two-color counters on ten frames*

https://qrs.ly/o2g99kg

Addition

Addition is the "**putting together**" of two groups of objects and finding how many in all. The numbers being added together are called **addends**.

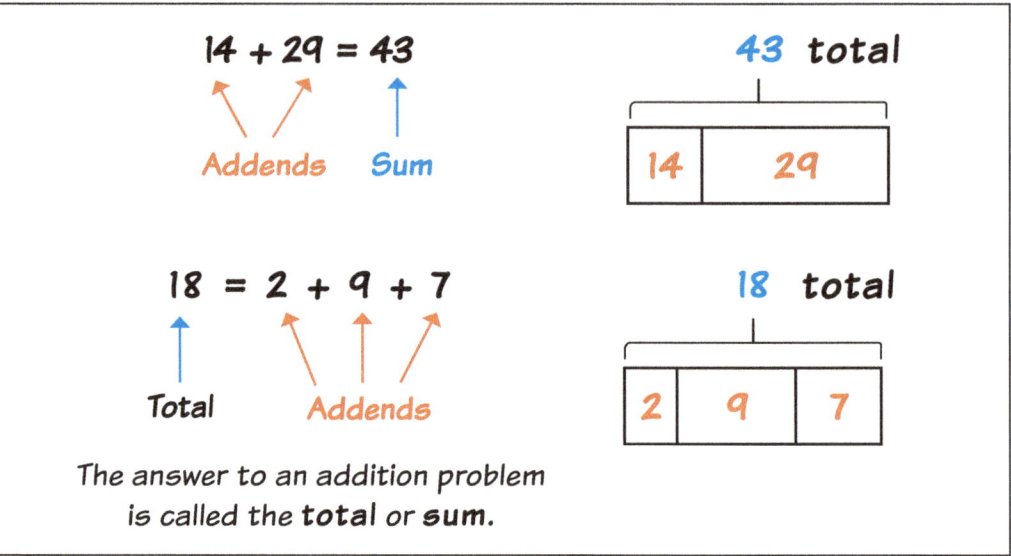

Subtraction

Subtraction is the "**taking apart**" or "**taking from**" a group of objects by removing one or more objects and finding the difference or "how many are left." Subtraction is also the **comparison** of two groups to find "**how many more or less**" are in each group.

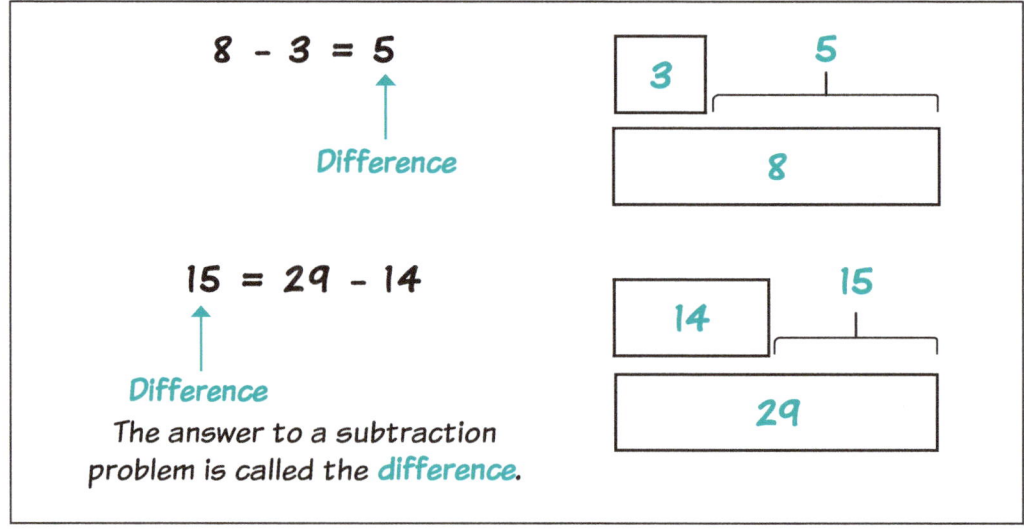

38   SEEING THE MATH YOU TEACH, GRADES K–6

The Relationship Between Addition and Subtraction . . . .

Addition and subtraction are **inverse operations**. That means they are opposite operations. They undo each other.

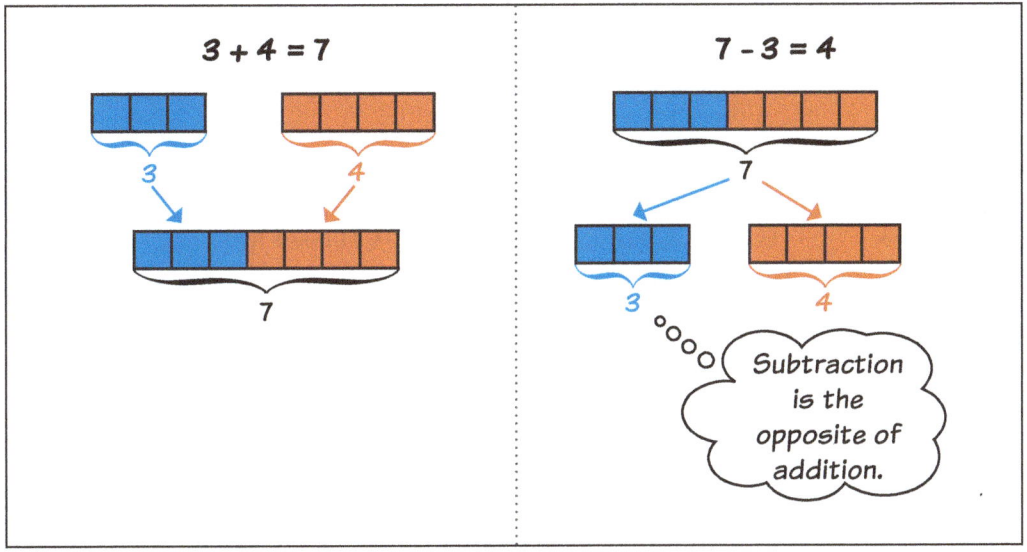

Fact Families

8 + 4 = 12   4 + 8 = 12

12 − 8 = 4   12 − 4 = 8

Since addition and subtraction are opposites, I can use the addition facts I know to help me solve subtraction problems!

Triangular Fact Cards

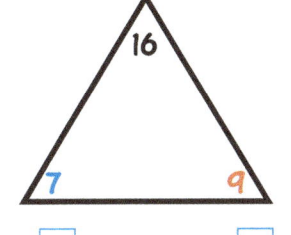

9 + [7] = 16   16 − [7] = 9

[9] + 7 = 16   16 − [9] = 7

I can study both addition and subtraction facts using a triangular fact card! I can cover up any number in the triangle and use my number facts to identify that missing number!

Addition and Subtraction Using Place Value

Writing Equations for Addition and Subtraction Word Problems....................

A word problem can have the **variable** or unknown number in different places.

Addition and Subtraction

	Result Unknown	Start Unknown	Change Unknown
Add To	5 bunnies sat on the grass. 7 more bunnies hopped over. How many bunnies are on the grass now? [5 | 7] with ? total 5 + 7 = ☐	Some bunnies were sitting on the grass. 7 more bunnies hopped over. Then there were 12 bunnies. How many bunnies were on the grass before? [? | 7] with 12 total ☐ + 7 = 12	5 bunnies were sitting on the grass. Some more bunnies hopped over. Now there are 12 bunnies. How many bunnies hopped over? [5 | ?] with 12 total 5 + ☐ = 12
Take From	12 apples were on the table. I ate 5 apples. How many apples are on the table now? 12 total; [5 | ?] 12 − 5 = ☐	Some apples were on the table. I ate 5 apples. Then there were 7 apples. How many apples were there before? ? total; [5 | 7] ☐ − 5 = 7	12 apples were on the table. I ate some of them, leaving 7 apples. How many apples did I eat? 12 total; [? | 7] 12 − ☐ = 7

Writing Equations for Addition and Subtraction Word Problems (continued)

A word problem can sometimes have more than one **variable** or unknown number in different places.

	Total Unknown	Addend Unknown	Both Addends Unknown*
Put Together/Take Apart	7 pink balls and 5 green balls are in the toy box. How many balls total are in the toy box? ? over [7][5] 7 + 5 = ☐	12 balls are in the toy box. 7 are pink and the rest are green. How many of the balls in the toy box are green? 12 over [7][?] 12 = 7 + ☐ 12 − 7 = ☐	Grandma has 7 flowers and 2 vases. How many flowers can she put in her green vase, and how many can she put in her pink vase? 7 over [?][?] 7 = ☐ + ☐ *See page 42 for more about word problems with multiple answers.
Compare	**Difference Unknown** Collin has 5 carrots. Brian has 7 carrots. How many more carrots does Brian have than Collin? [5][?] [7] 5 + ☐ = 7 7 − 5 = ☐	**Bigger Unknown** Brian has 2 more carrots than Collin. Collin has 5 carrots. How many carrots does Brian have? ? Brian [ ][2] Collin [5] 2 + 5 = ☐ 5 + 2 = ☐	**Smaller Unknown** Brian has 2 more carrots than Collin. Brian has 7 carrots. How many carrots does Collin have? 7 Brian [ ][2] Collin [?] 7 − 2 = ☐ 2 + ☐ = 7

Addition and Subtraction Using Place Value

Addition and Subtraction Word Problems.......

There are four common types of addition and subtraction word problems.

Add To

6 birds sat on a fence. 3 more birds flew there. How many birds are on the fence now?

6 + 3 = ?

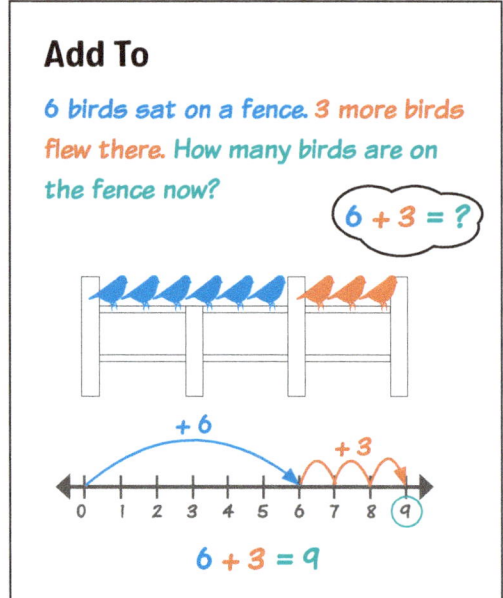

6 + 3 = 9

Take From

9 oranges were in a bowl. I ate some oranges. Then there were 6 oranges. How many oranges did I eat?

9 − ? = 6

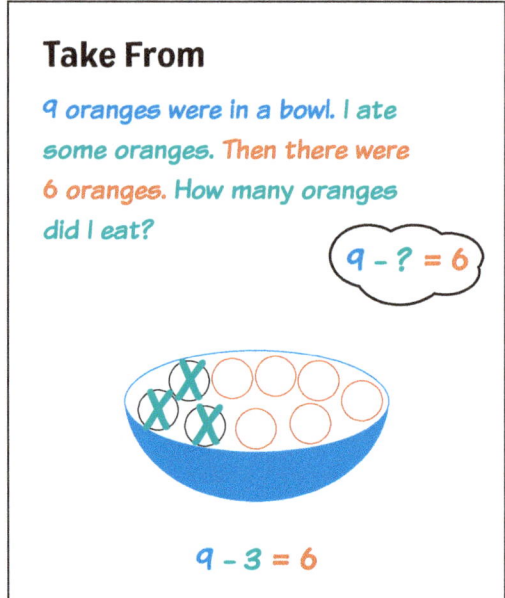

9 − 3 = 6

Put Together/Take Apart

9 dinner rolls are on the table. 6 are wheat, and the rest are white. How many dinner rolls are white?

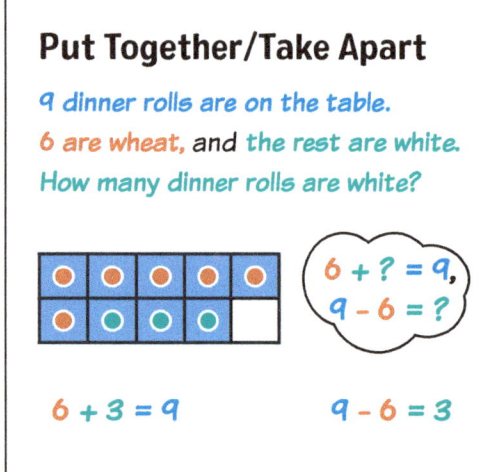

6 + ? = 9,
9 − 6 = ?

6 + 3 = 9   9 − 6 = 3

Compare

Sam has 3 fewer crayons than Tim. Tim has 9 crayons. How many crayons does Sam have?

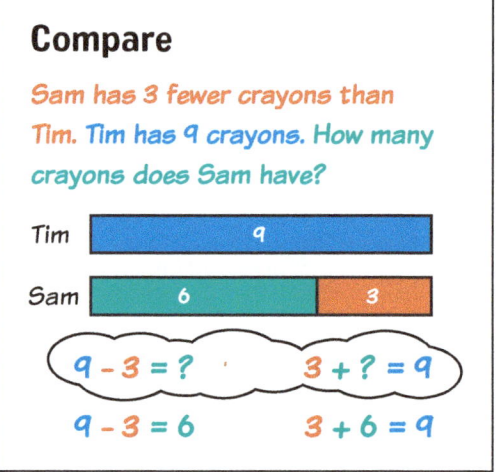

9 − 3 = ?   3 + ? = 9

9 − 3 = 6   3 + 6 = 9

 Sometimes problems have many possible answers.

Taylor has 9 marbles. Some are red and some are yellow. How many of each color marble could Taylor have?

9 = ? + ?

1 red + 8 yellow   8 red + 1 yellow
2 red + 7 yellow   7 red + 2 yellow
3 red + 6 yellow   6 red + 3 yellow
4 red + 5 yellow   5 red + 4 yellow

These are all possible solutions because they each equal a total of 9 marbles.

Addition and Subtraction
Two-Step Word Problems

Two-step word problems take two separate actions to solve.

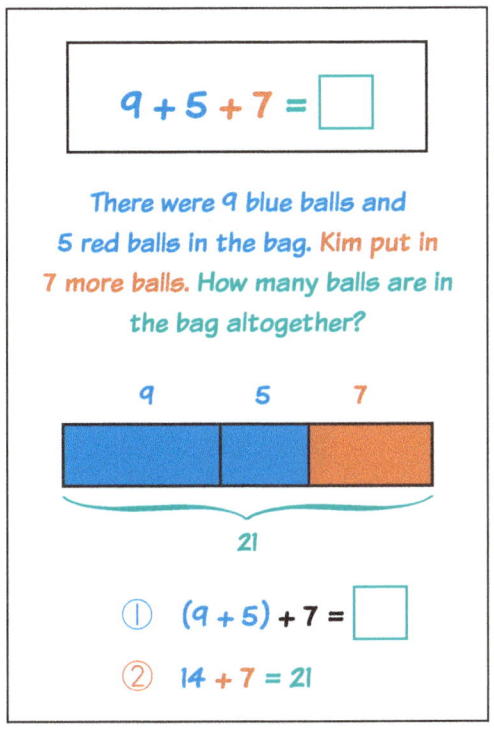

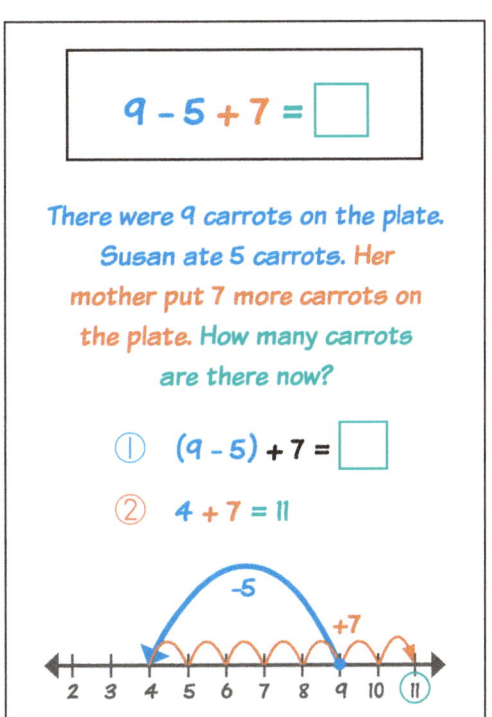

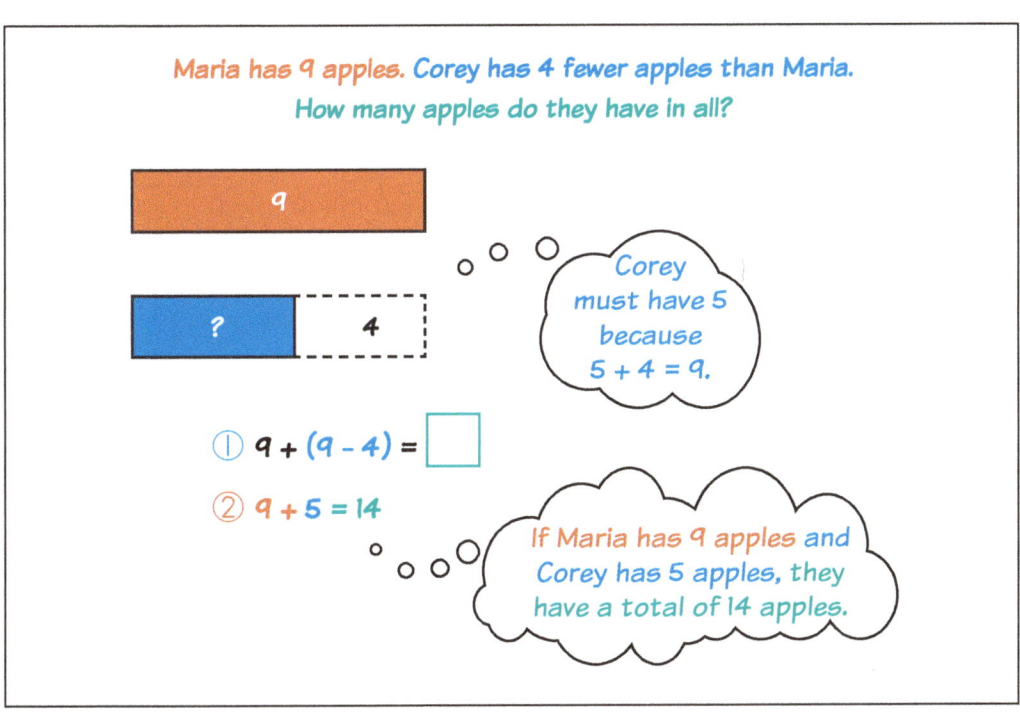

Addition and Subtraction Using Place Value

Adding Whole Numbers

There are many methods for **adding whole numbers**.

Counting On

7 + 2   Start with 7 and count on 2 more.

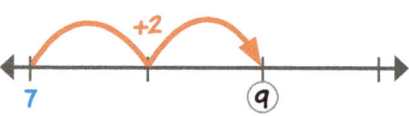

14 + 5   Start with 14 and count on 5 more.

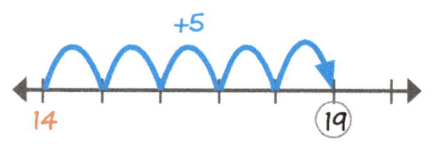

Using Known Sums

6 + 7
= (6 + 6) + 1
= 12 + 1
= 13

I know 6 + 6 = 12, so 1 more would be 13.

4 + 11 + 8
= 4 + 3 + (8 + 8)
= 4 + 3 + 16
= 3 + (4 + 16)
= 3 + 20
= 23

I know 8 + 8 = 16, so I decomposed the 11 into 3 & 8.

I know 16 + 4 = 20, so I rearranged the numbers to make 20, then added 3.

Making 10

If you know your combinations of 10, you can combine addends and make the problem friendly.

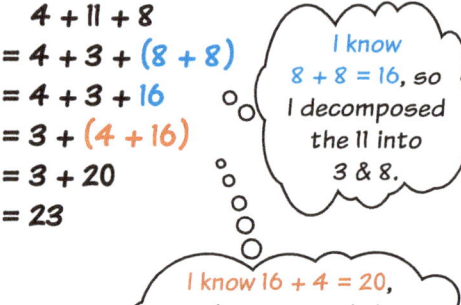

○ = 8 + (2 + 4)   = (16 + 1) + 9
○ = (8 + 2) + 4   = 16 + (1 + 9)
○ = 10 + 4   = 16 + 10
○ = 14   = 26

I'll decompose (break apart) the 6 into 2 + 4. That way I can "make ten" out of the 8 + 2!

Number Pairs That Total 10

0 + 10   10 + 0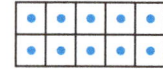

1 + 9   9 + 1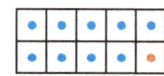

2 + 8   8 + 2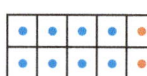

3 + 7   7 + 3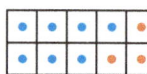

4 + 6   6 + 4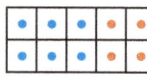

5 + 5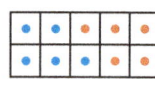

Adding Larger Whole Numbers

Numbers can be added by paying attention to **place value**.

Video 4: Addition With Place-Value Pieces: You can use place value pieces to help students "see" the underlying concepts in the standard algorithm for addition. *Pictured here: base-ten blocks*

https://qrs.ly/1lg99jz

Combining 100s, 10s and 1s

```
      456
    + 167
```
① First, add the hundreds.   ④ { 500   (400 + 100)   ①
② Second, add the tens.   110   (50 + 60)   ②
③ Third, add the ones.   +   13   (6 + 7)   ③
④ Fourth, add them all together.   623

```
    1,298
    + 973
```
① First, add the thousands.   ⑤ { 1,000   (1,000 + 0)   ①
② Second, add the hundreds.   1,100   (200 + 900)   ②
③ Third, add the tens.   160   (90 + 70)   ③
④ Fourth, add the ones.   +   11   (8 + 3)   ④
⑤ Fifth, add them all together.   2,271

Traditional Method

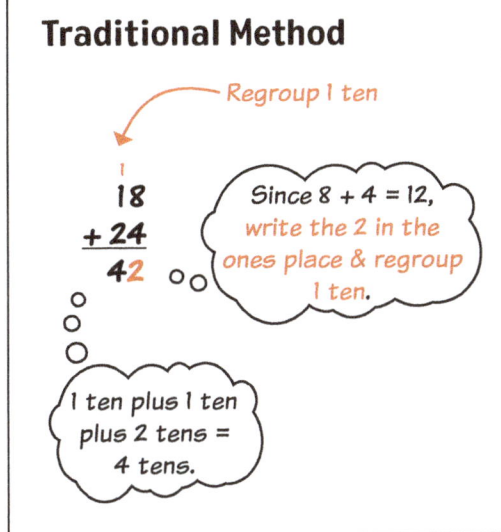

Since 8 + 4 = 12, write the 2 in the ones place & regroup 1 ten.

1 ten plus 1 ten plus 2 tens = 4 tens.

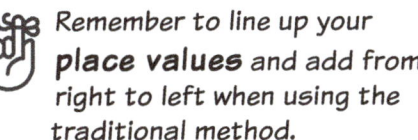 Remember to line up your **place values** and add from right to left when using the traditional method.

Traditional Method

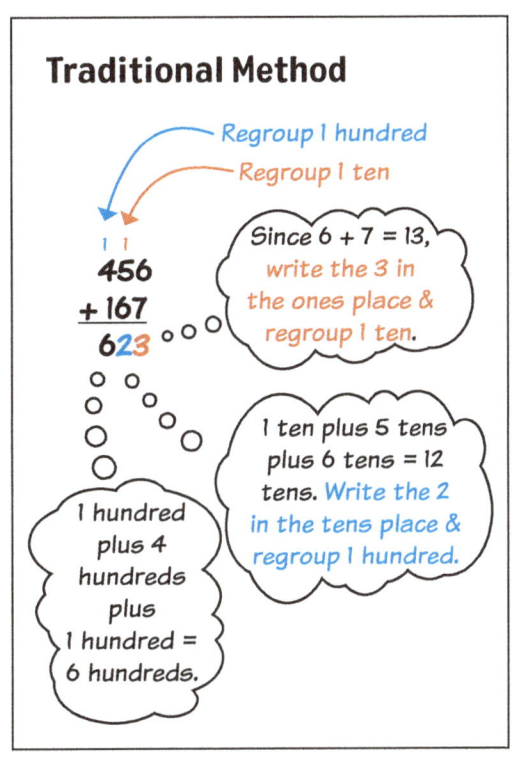

Since 6 + 7 = 13, write the 3 in the ones place & regroup 1 ten.

1 hundred plus 4 hundreds plus 1 hundred = 6 hundreds.

1 ten plus 5 tens plus 6 tens = 12 tens. Write the 2 in the tens place & regroup 1 hundred.

Addition and Subtraction Using Place Value

Subtracting Using the Adding-On Method.......

You can use addition to help you subtract both large and small numbers.

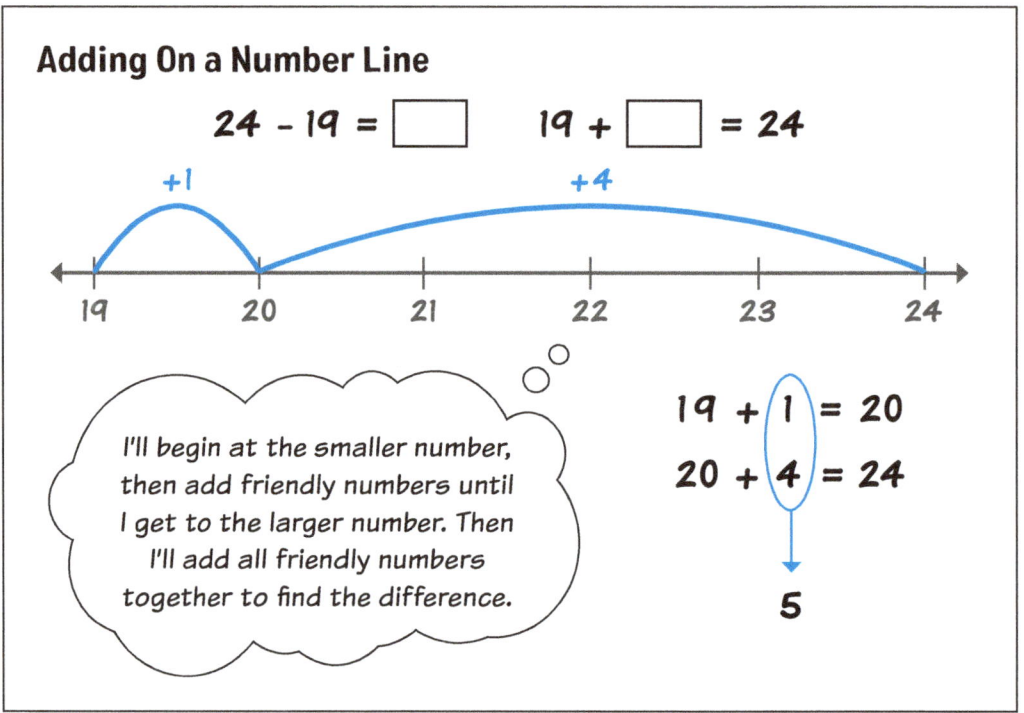

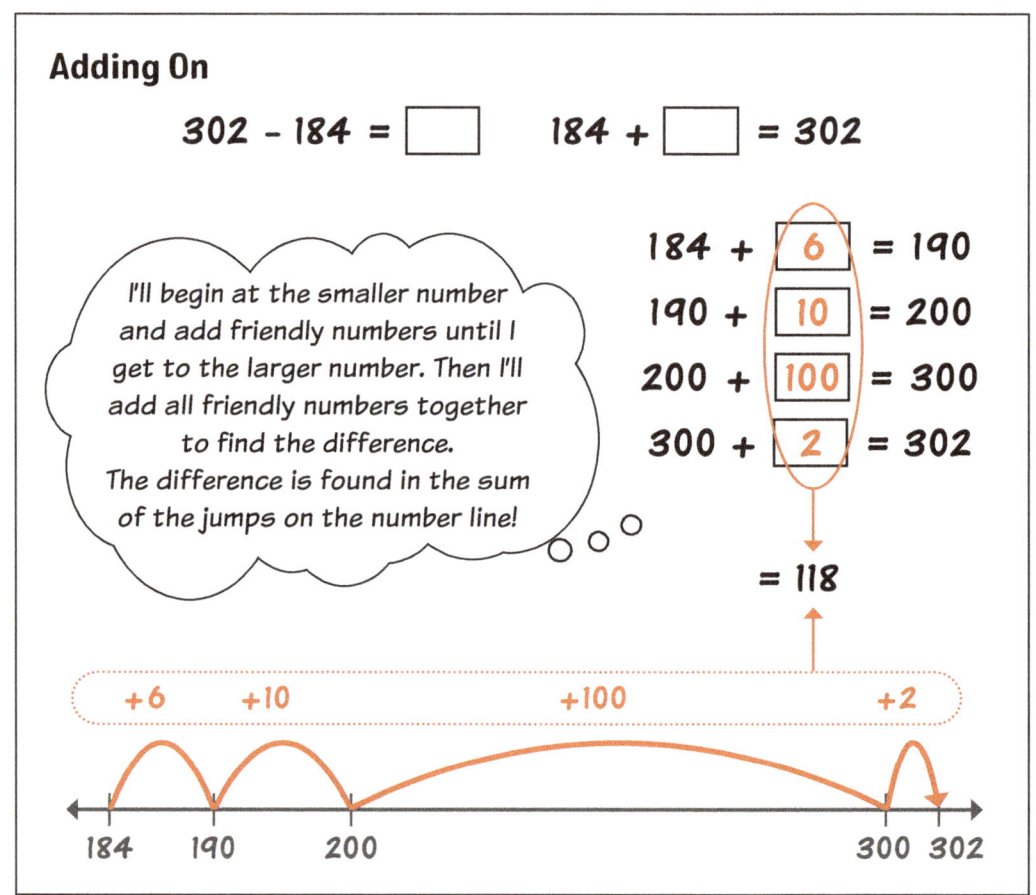

Subtracting Whole Numbers

There are many methods for **subtracting whole numbers**. Some methods work better for smaller numbers and some work better for larger numbers.

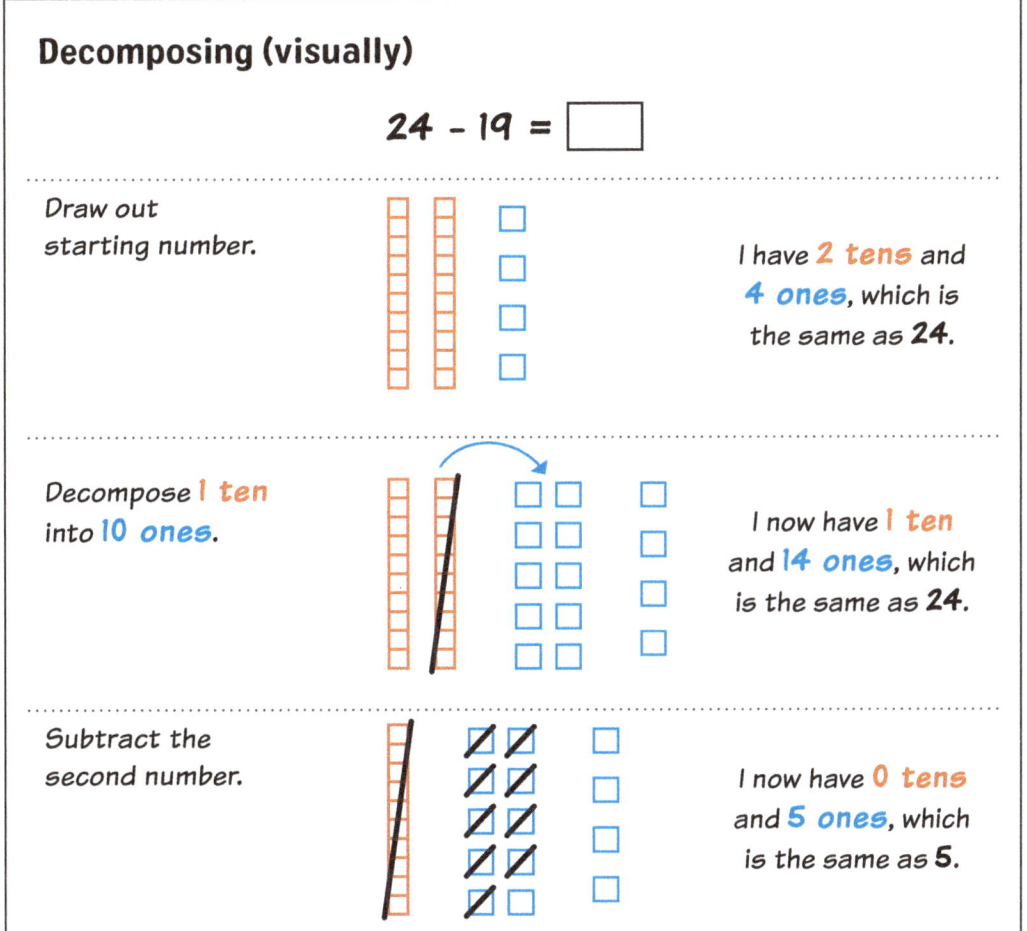

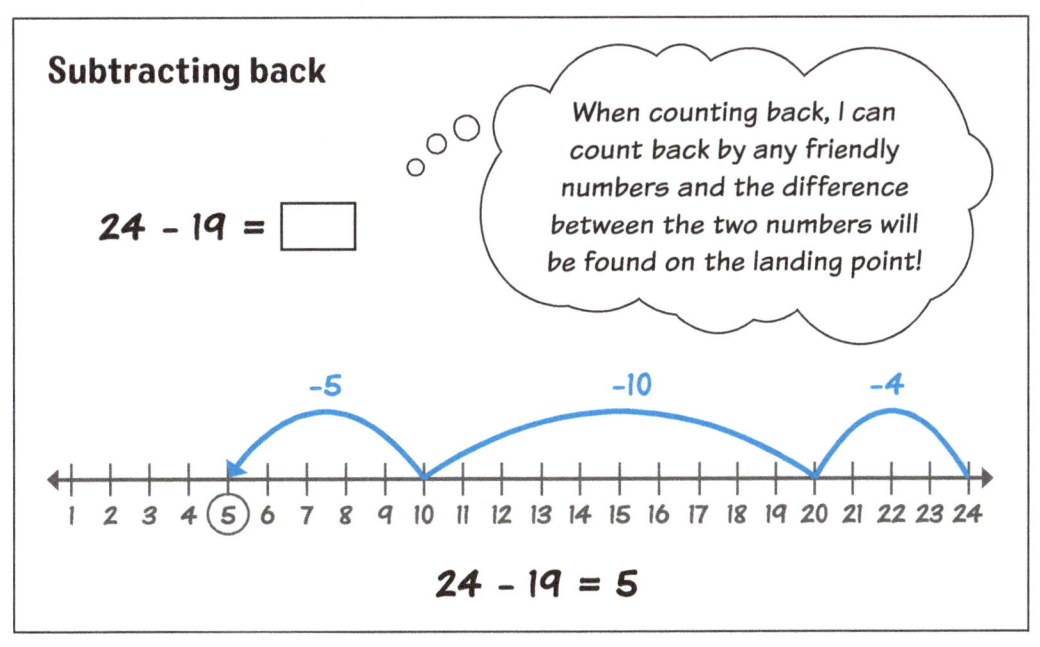

Addition and Subtraction Using Place Value

Subtracting Whole Numbers (continued)

There are many methods for **subtracting whole numbers**. Some methods work better for smaller numbers and some work better for larger numbers.

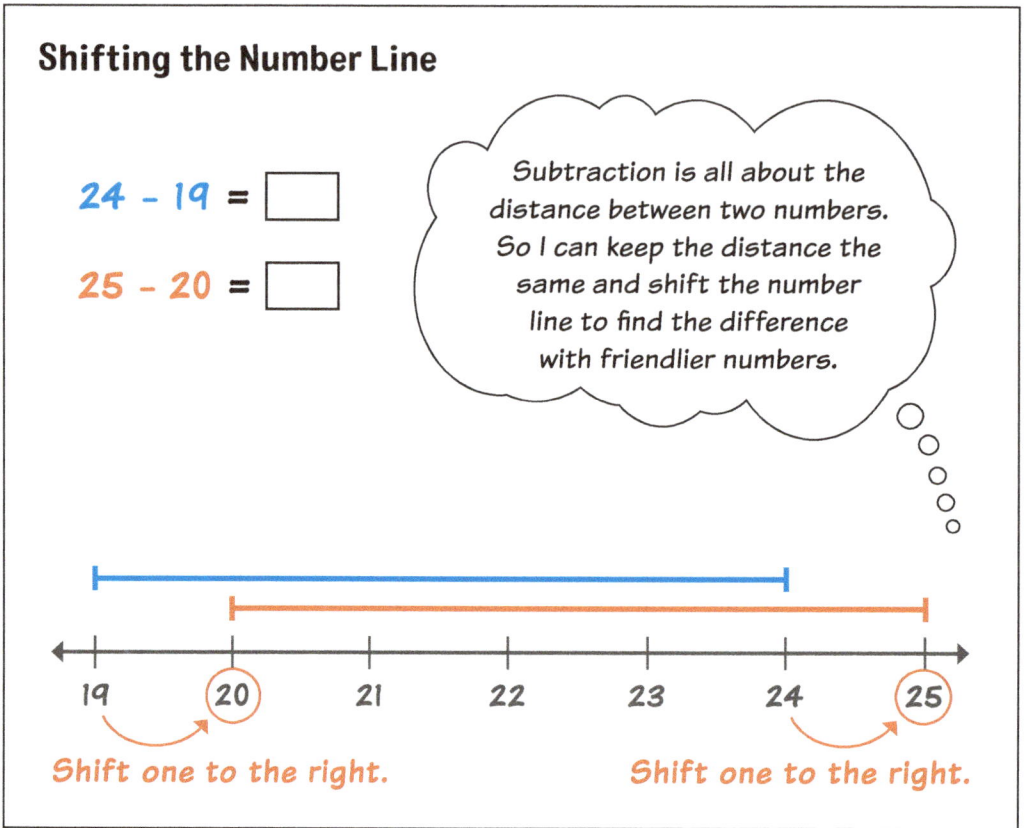

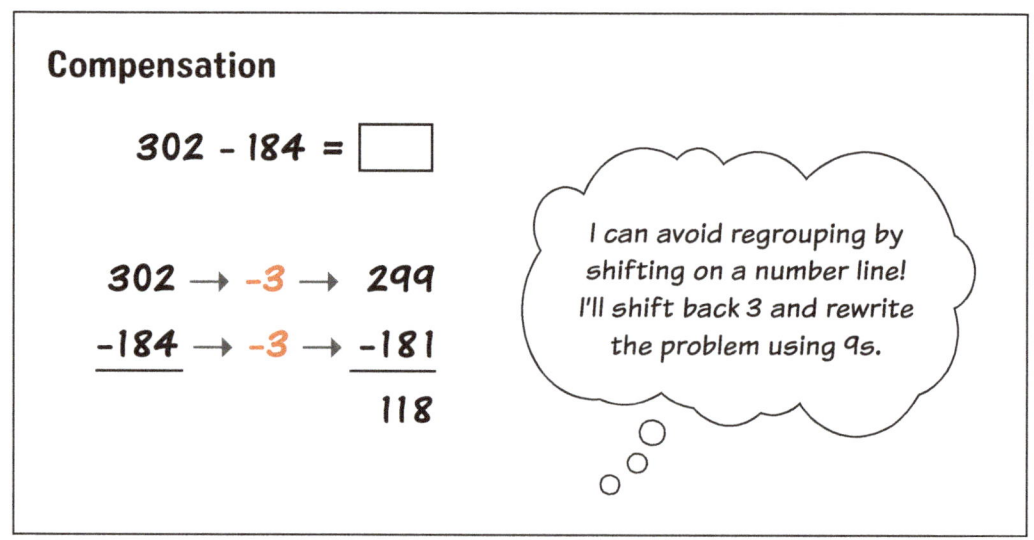

Subtracting Larger Whole Numbers

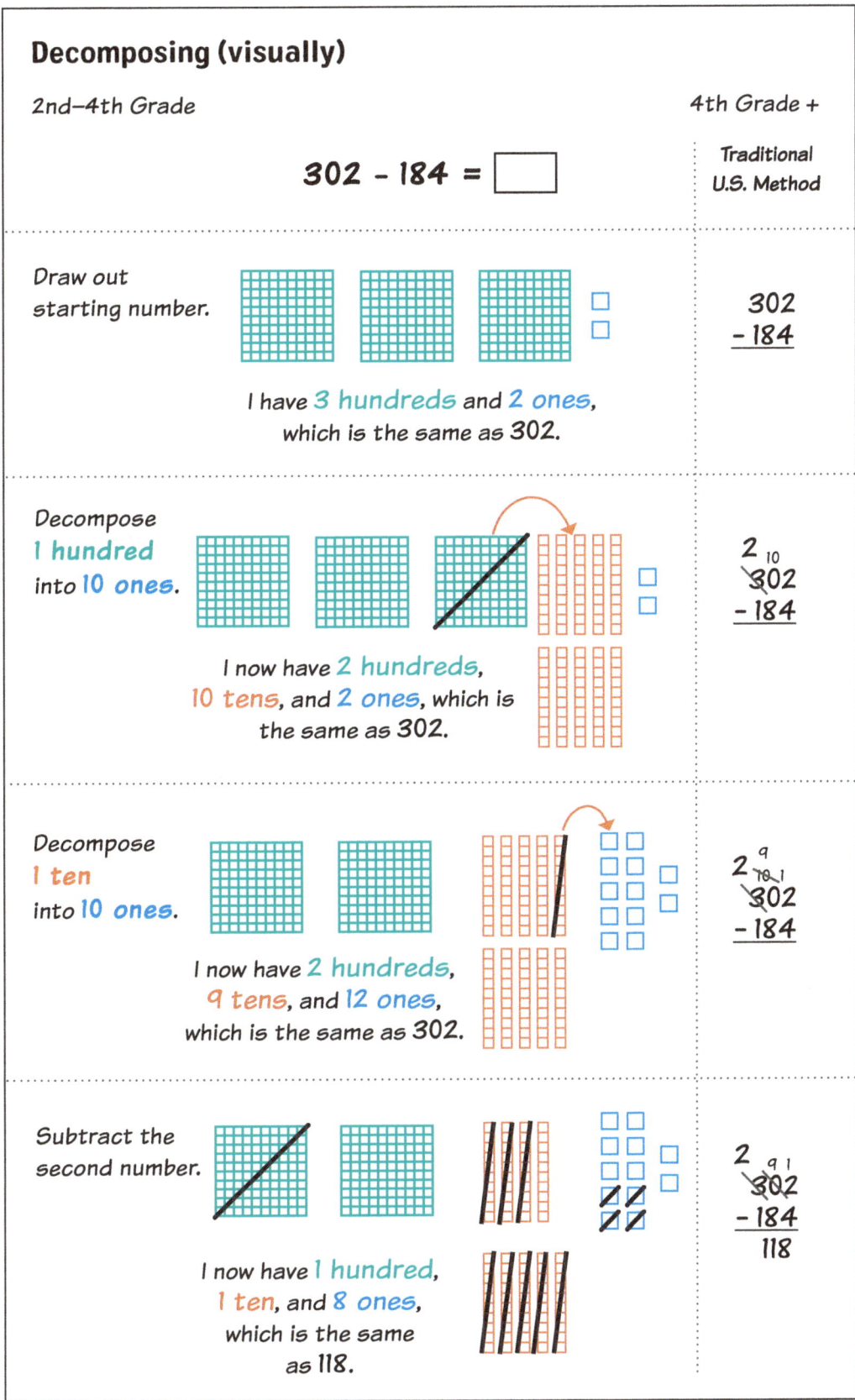

Addition and Subtraction Using Place Value

Subtracting Using the Traditional Method......

Here are a few examples of subtraction using the **traditional U.S. method**. The traditional U.S. method begins to be used in fourth grade. Just like in addition, place values (columns) must be "lined up" so regrouping is easier.

Video 5: Subtraction With Place-Value Pieces: You can use place value pieces to help students "see" the underlying concepts in the standard algorithm for subtraction. *Pictured here: place value disks*

https://qrs.ly/2rg99kj

No Regrouping Needed

$$\begin{array}{r} 2{,}984 \\ -612 \\ \hline 2{,}372 \end{array}$$

- Subtract each column, starting with the ones and moving left.

Some Regrouping Needed

$$\begin{array}{r} {}^{3}\,{}^{16} \\ 14\cancel{6} \\ -38 \\ \hline 108 \end{array}$$

- Subtract the ones...not enough! You need more!
- Use 1 ten. Since 1 ten = 10 ones, change the tens to 3 and add 10 to the value in the ones place.
- Now subtract the ones.

Lots of Regrouping Needed

$$\begin{array}{r} {}^{13}\\ {}^{9\,3\,13}\\ \cancel{1}{,}04\cancel{3} \\ -754 \\ \hline 289 \end{array}$$

- Subtract the ones—not enough! Use 1 ten, changing 4 tens to 3 tens and the 3 in the ones place to 13 ones. Now subtract.
- Subtract the tens—not enough! Oh my...there is a zero in the hundreds place. Move on to the thousands place. Since 1 thousand = 10 hundreds, use 1 hundred (leaving 9 hundred in the hundreds place), changing the 3 in the tens place to 13 tens. Now subtract.
- Subtract the hundreds.

 The U.S. traditional method for subtraction uses decomposition written in shorthand.

Adding and Subtracting Decimals..............

Adding and subtracting decimals is a lot like adding and subtracting whole numbers. We add or subtract like units with each other (ones to ones, tenths to tenths, and so on). The decimal points that separate the whole number parts from the fractional parts are "lined up" to help keep the place values together (tens, ones, tenths, etc.).

Adding Decimals

Christian was hiking in the Grand Canyon. His watch indicated he had hiked 5.213 km on one trail and 3.3 km on another trail before he ate lunch. How many kilometers did Christian hike before lunch?

5.213 + 3.3 = ☐

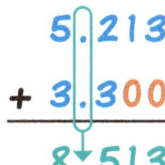

- Write the numbers, one under the other with the decimal points lined up.
- Insert zeroes so the numbers have the same amount of places.
- Add normally and bring the decimal point straight down into the answer.

Subtracting Decimals

Collin was hiking in the Grand Canyon. He was hiking a 5.213-km trail. After some time, he saw a sign that read, "3.3 km to trail's end." How many kilometers had Collin hiked already?

5.213 - 3.3 = ☐

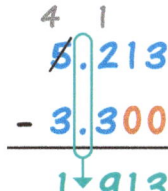

- Write the numbers, one under the other with the decimal points lined up.
- Insert zeroes so the numbers have the same amount of places.
- Subtract normally and bring the decimal point straight down into the answer.

Addition and Subtraction Using Place Value

Adding and Subtracting Decimals (continued)

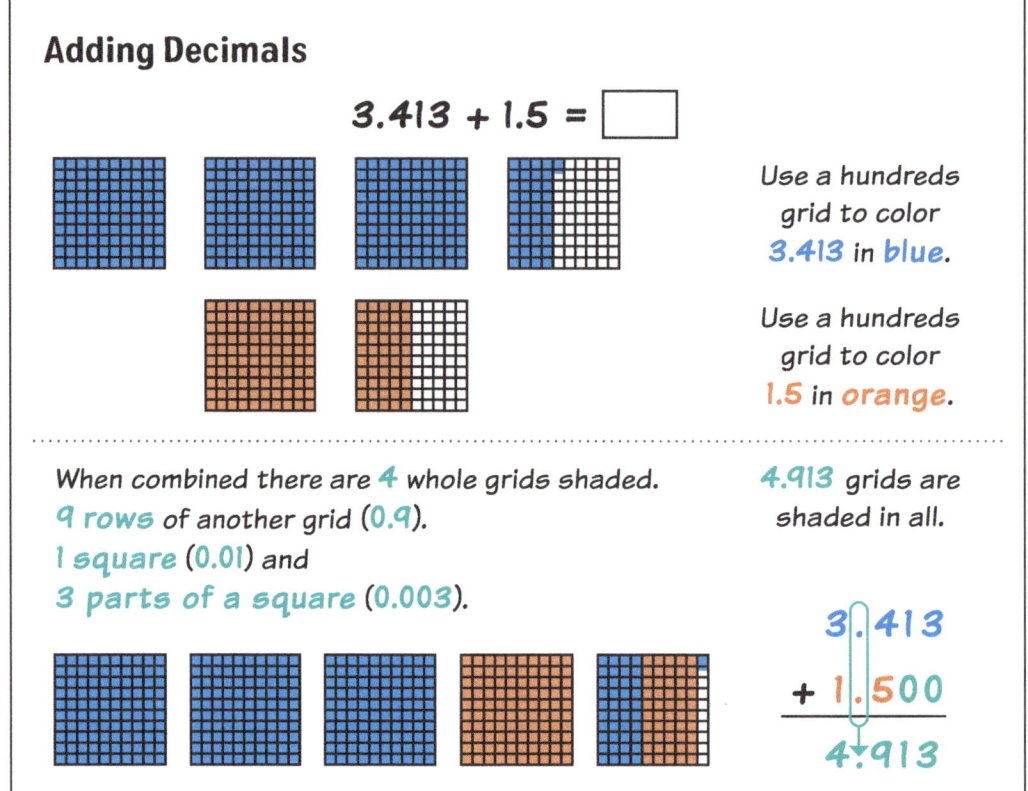

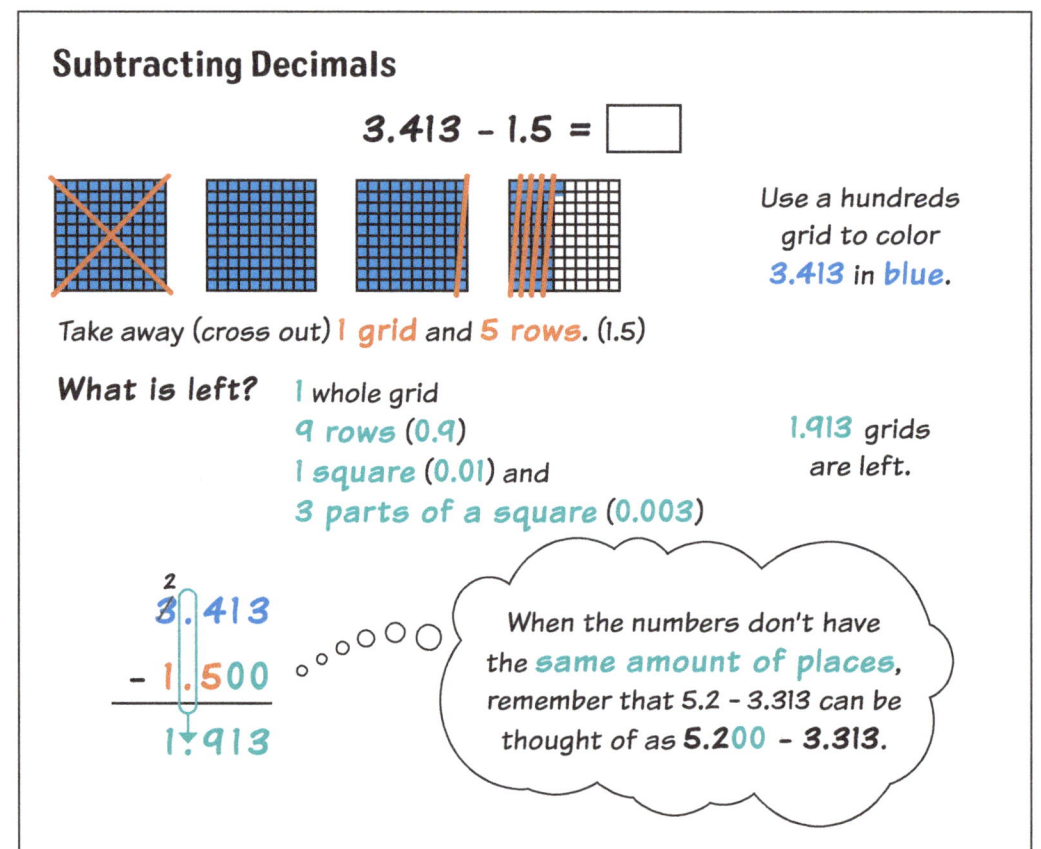

Chapter 5

Multiplication and Division Using Place Value

Multiplication and division are often taught close together because they are inverse operations. This means that the operations are opposites and can "undo" each other. For example, multiplication is initially introduced as joining equal groups to find how many there are in total, but this operation can be "undone" by separating or dividing the total into equal groups.

When students understand this inverse relationship, they can think of division problems as finding the unknown factor, interpreting a problem like $10 \div 5 = ?$ as if it were $5 \times ? = 10$. Because division is as unnatural as multiplication is natural, students may find it easier to think of division as a missing factor question. Furthermore, when students understand this inverse relationship between multiplication and division, they become stronger problem solvers.

Similarly, in later grades, students think of multiplying as scaling. Scaling can be seen in the real world when a photograph is enlarged to twice its size to make a poster (multiplication). This action can be undone by scaling a large painting to half of its size (division).

It's important to note that, unlike addition, multiplication has several interpretations. The most agreed-upon interpretation includes equal groups, area, multiplicative comparison, rates, and Cartesian products. This notion also impacts the ways in which we think about division.

Typical Trajectory in Most State Standards Frameworks:

- Grades K–2: Pre-multiplication activities such as skip counting, repeated addition, and repeated addition using an array model
- Grade 3: Multiplication through 10 × 10 and the related division facts
- Grade 4: Multi-digit multiplication using conceptual strategies; multi-digit division with one-digit divisors using conceptual strategies
- Grade 5: Multi-digit multiplication using a standard algorithm; multi-digit division with two-digit divisors using conceptual strategies; multi-digit multiplication and division with decimal numbers using conceptual strategies
- Grade 6: Multi-digit whole-number and decimal multiplication and division using standard algorithms

Multiplication

Multiplication is basically the combining of equal sized groups to find how many in all. The numbers being multiplied are called **factors**.

Video 6: Factors With Linking Cubes: You can use linking cubes to help students "see" the distinct behaviors of the two factors in a multiplication expression. Pictured here: Unifix cubes

https://qrs.ly/6fg99kl

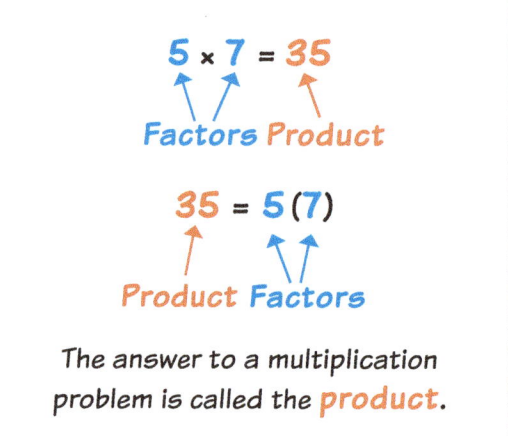

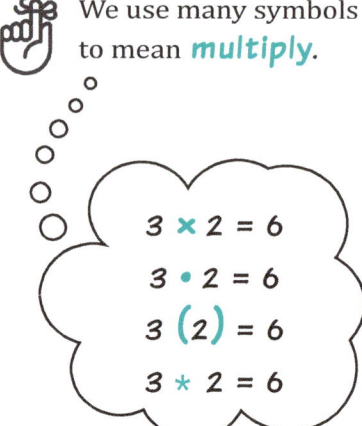

Division

Division is basically the splitting of an object or set into equal parts or groups. The set is called the **dividend**, and the number of parts or groups is called the **divisor**. Division can be represented using a fraction or an equation.

56   SEEING THE MATH YOU TEACH, GRADES K–6

The Relationship Between Multiplication and Division

Multiplication and division are **inverse operations**, which means they are opposite operations. They undo each other.

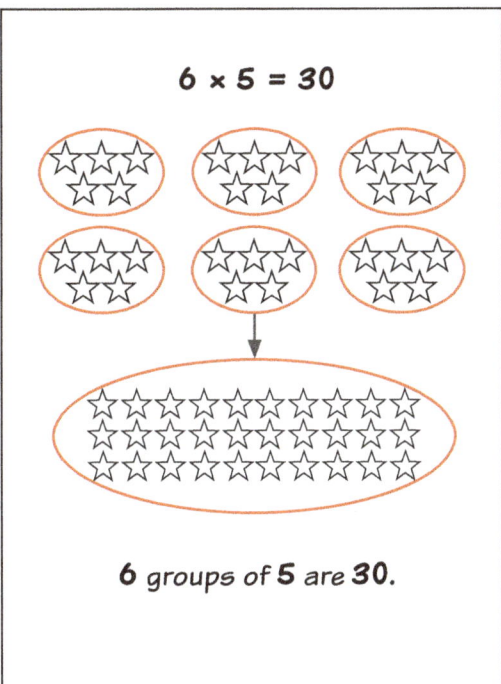

6 × 5 = 30

6 groups of 5 are 30.

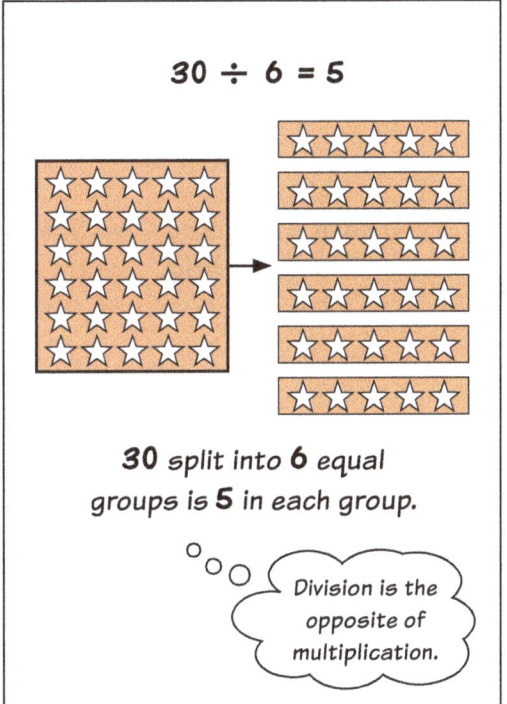

30 ÷ 6 = 5

30 split into 6 equal groups is 5 in each group.

Division is the opposite of multiplication.

Fact Families

A fact family is a group of all the related multiplication and division facts between three related numbers.

8 × 7 = 56
56 ÷ 8 = 7
7 × 8 = 56
56 ÷ 7 = 8

I can use the multiplication facts I know to help me solve division problems!

Triangle Facts

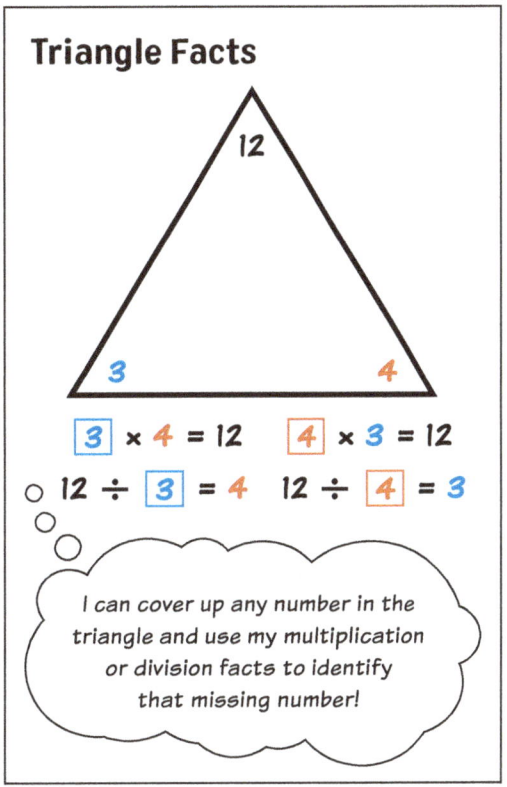

3 × 4 = 12   4 × 3 = 12
12 ÷ 3 = 4   12 ÷ 4 = 3

I can cover up any number in the triangle and use my multiplication or division facts to identify that missing number!

Multiplication and Division Using Place Value

Writing Equations for Multiplication and Division Word Problems...........

A multiplication or division word problem can have the **variable** or unknown number in different places.

Product Unknown

$a \times b = \square$

There are 7 bags with 3 plums in each bag. How many plums are there in all?

$7 \times 3 = \square$

| 3 | 3 | 3 | 3 | 3 | 3 | 3 |

?

Group Size Unknown

$a \times \square = c$

$c \div a = \square$

If 21 plums are shared equally in 7 bags, how many plums are in each bag?

$7 \times \square = 21$

$21 \div 7 = \square$

21

?

Number of Groups Unknown

$\square \times b = c$

$c \div b = \square$

If 21 plums are packed 3 to each bag, how many bags are needed?

$\square \times 3 = 21$

$21 \div 3 = \square$

21

| 3 | 3 | 3 | 3 | 3 | 3 | 3 |

How many bags?

Multiplication Word Problems

There are three major types of problems related to **multiplication**.

Equal Groups

Josie has 3 toy cars. Each car has 4 wheels. How many wheels are there in all?

3 groups of 4

3 × 4 = 12

Arrays of Objects

At the toy store, Kaitlin notices 3 rows of dolls with 4 dolls in each row. How many dolls are at the toy store?

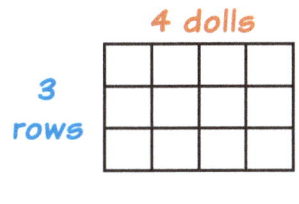

3 × 4 = 12

Compare

A small balloon costs $3.00. The large balloon costs 4 times as much. How much does the large balloon cost?

$3 × 4 = $12

When you solve a "**compare**" problem in multiplication, you are using multiplication as a **scaling factor**.

Multiplication allows you to compute multiple copies of the same-sized group without having to add repeatedly.

A store has 27 boxes of granola bars. There are 9 bars in each box. How many granola bars are there in all?

Rather than add 27 together 9 times, you can **multiply**!

27 × 9 = ☐

27 × 9 = 243

When thinking about 3 × 9 and 9 × 3 it's important to know that even though they have the same product, they are conceptually different.
   3 × 9 can be thought of as having 3 bags with 9 cookies each, whereas 9 × 3 is like having 9 bags with 3 cookies each. It's still 27 cookies, but what a different experience!

Multiplication and Division Using Place Value

Multiplication as a Scale Factor

In upper grades, multiplication expressions can be interpreted in terms of a quantity and a **scaling factor**. A scaling factor is a number which scales, or multiplies some quantity. When scaling you are typically finding how many times larger or smaller one size is when compared to another.

Video 7: Multiplicative Comparison With Linking Cubes and Ten Frames: You can use linking cubes and ten frames to help students compare and contrast the concepts of additive comparison and multiplicative comparison. *Pictured here: Unifix cubes and KP® Ten-Frame Tiles*

https://qrs.ly/v9g99ko

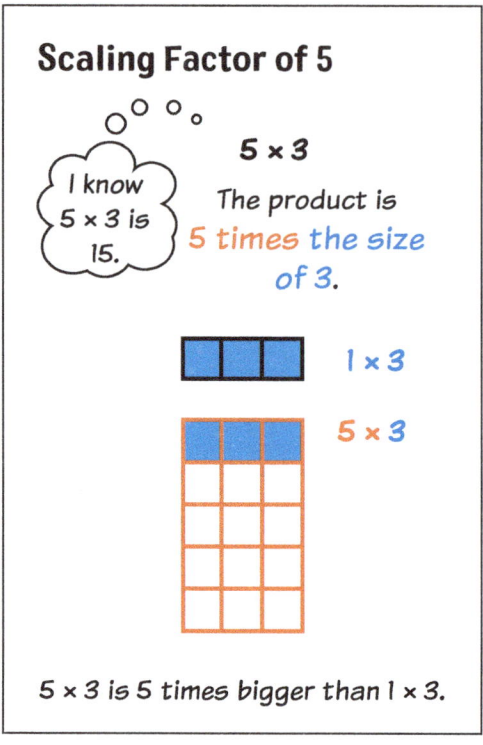

Scaling Factor of 5

I know 5 × 3 is 15.

5 × 3

The product is **5 times the size of 3.**

1 × 3

5 × 3

5 × 3 is 5 times bigger than 1 × 3.

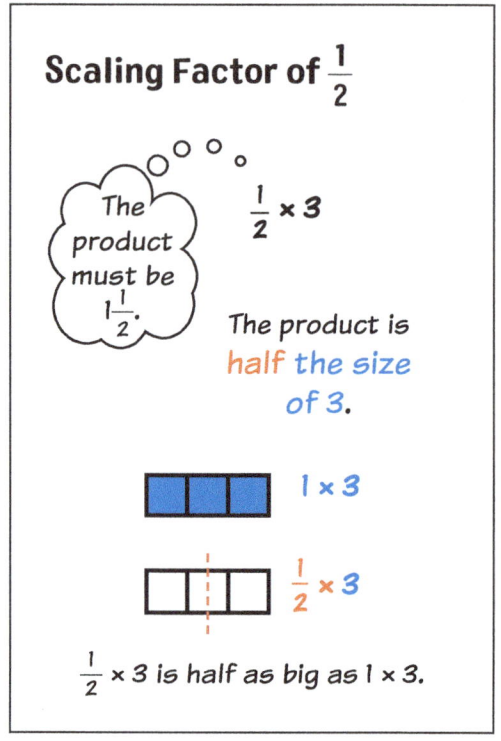

Scaling Factor of $\frac{1}{2}$

The product must be $1\frac{1}{2}$.

$\frac{1}{2}$ × 3

The product is **half the size of 3.**

1 × 3

$\frac{1}{2}$ × 3

$\frac{1}{2}$ × 3 is half as big as 1 × 3.

Gary threw his ball 5 feet. Marie threw her ball three times farther than Gary threw his ball.

How far did Marie throw her ball?

3 × 5 feet = 15 feet
5 feet

Gary's Ball   Marie's Ball

Marie threw her ball 15 feet.

Three friends are comparing their collections of toy cars. JD has 24 cars, Jose has 9 cars, and Bill has twice as many toy cars as both of his friends combined.

How many cars does Bill have?

Bill's Cars = 2 (24 + 9)

Twice as many   JD's and Jose's cars

2 (33) = 66

Bill has 66 toy cars.

Multiplication Table

A multiplication table can be used to help you quickly identify your multiplication facts.

×	1	2	3	4	5	6	7	8	9	10	11	12
1	1	2	3	4	5	6	7	8	9	10	11	12
2	2	4	6	8	10	12	14	16	18	20	22	24
3	3	6	9	12	15	18	21	24	27	30	33	36
4	4	8	12	16	20	24	28	32	36	40	44	48
5	5	10	15	20	25	30	35	40	45	50	55	60
6	6	12	18	24	30	36	42	48	54	60	66	72
7	7	14	21	28	35	42	49	56	63	70	77	84
8	8	16	24	32	40	48	56	64	72	80	88	96
9	9	18	27	36	45	54	63	72	81	90	99	108
10	10	20	30	40	50	60	70	80	90	100	110	120
11	11	22	33	44	55	66	77	88	99	110	121	132
12	12	24	36	48	60	72	84	96	108	120	132	144

To find the product of 7 × 8, you can

- go to row seven and over to column eight.
- go to row eight and over to column seven.

Since 8 × 7 = 7 × 8 both products are 56.

Try it with another fact: 6 × 4

- row six, column four

    6 × 4 = 24

- row four, column six

    4 × 6 = 24

 You can use the multiplication table to notice patterns. For example, *all multiples of 4 are even*, and *all multiples of 10 have a 0 in the ones place*.

Multiplication and Division Using Place Value

Single-Digit Multiplication Strategies

There are many different strategies for **multiplication**.

3×7

Skip Counting

Count by 3:

3, 6, 9, 12, 15, 18, (21)

I'll count by 3, 7 times.

Count by 7:

7, 14, (21)

I'll count by 7, 3 times.

Arrays

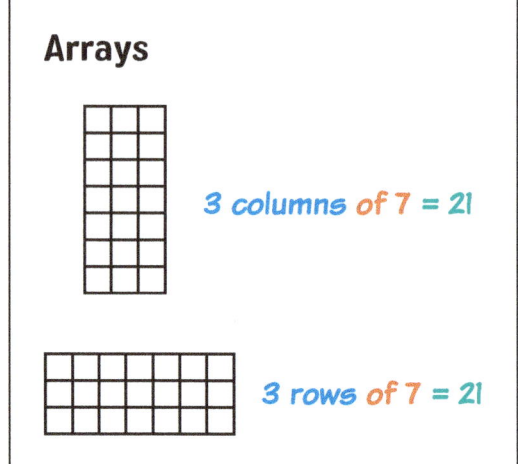

3 columns of 7 = 21

3 rows of 7 = 21

Equal Groups

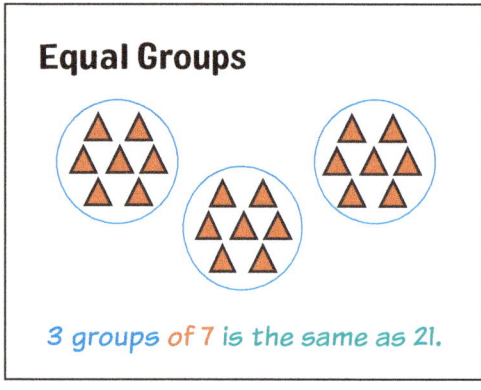

3 groups of 7 is the same as 21.

Area Model

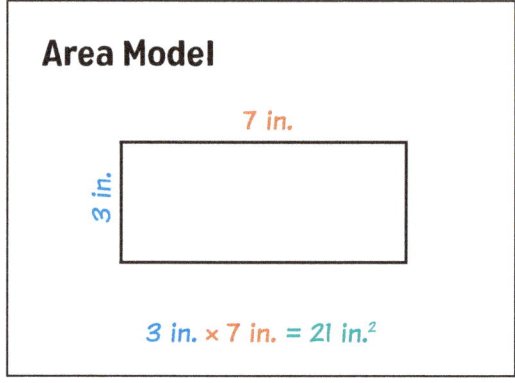

7 in.

3 in.

3 in. × 7 in. = 21 in.²

Decomposition/Distributive Property

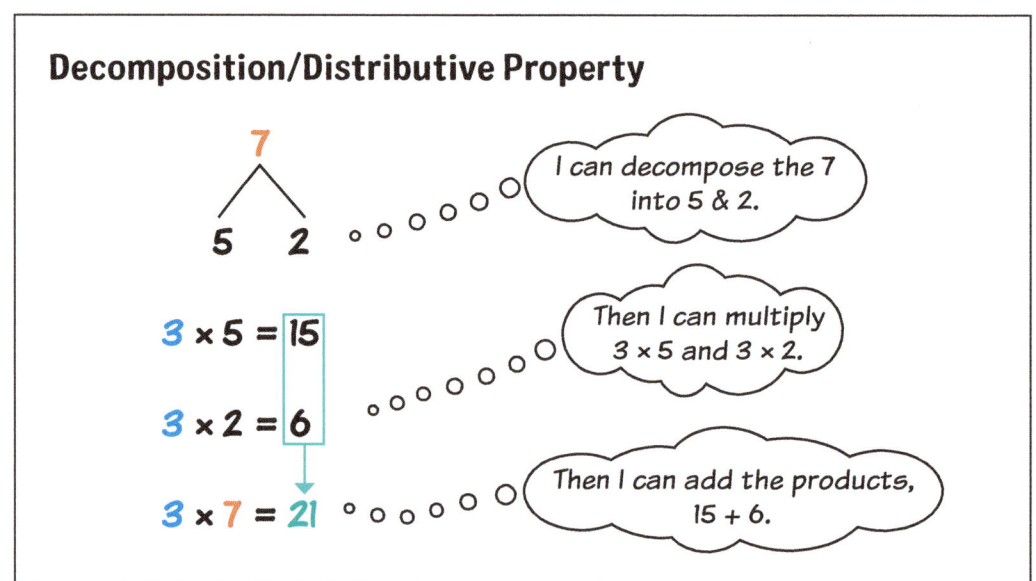

7
/ \
5   2

I can decompose the 7 into 5 & 2.

$3 \times 5 = 15$

Then I can multiply 3×5 and 3×2.

$3 \times 2 = 6$

$3 \times 7 = 21$

Then I can add the products, 15 + 6.

Multiplying by Multiples of Ten

When **multiplying numbers by multiples of ten**, you will see an example of the **associative property** of multiplication.

*The **associative property** states that changing the grouping of factors does not change the product.*

 How do you know 15 tens is 150?

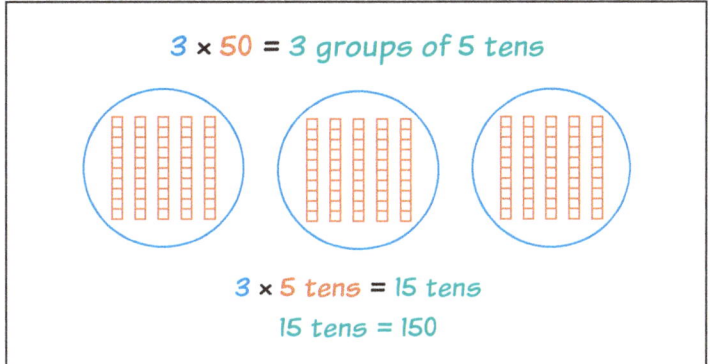

$3 \times 50 = 3$ groups of 5 tens

3×5 tens $= 15$ tens

15 tens $= 150$

Skip Counting by 50

5 tens is 50

10 tens is 100

15 tens is 150

Decomposing 15 tens

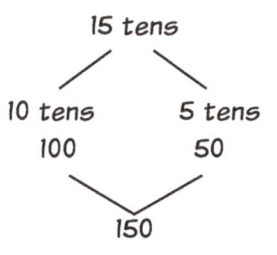

15 tens

10 tens   5 tens

100   50

150

Decomposing 15 × 10 Using Distributive Property

$15 \times 10 = (10 + 5) \times 10$
$= (10 \times 10) + (5 \times 10)$
$= 100 + 50$
$= 150$

 Shortcut: Calculate the product of the non-zero digits, then shift the product one place to the left to make the result 10 times as large.

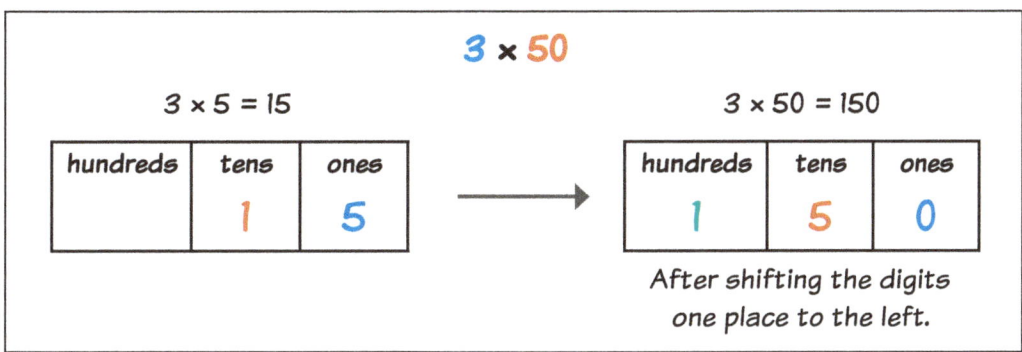

3×50

$3 \times 5 = 15$ $3 \times 50 = 150$

hundreds	tens	ones
	1	5

hundreds	tens	ones
1	5	0

After shifting the digits one place to the left.

Multiplication and Division Using Place Value

Multi-Digit Multiplication Strategies

There are many different strategies for **multiplication**. Here are a few:

13 × 5 = ☐

Skip Counting

Count by 13:

13, 26, 39, 52, (65)

○
○
○ *I'll count by 13, 5 times to get the product.*

Count by 5:

5, 10, 15, 20, 25, 30, 35, 40, 45, 50, 55, 60, (65)

○
○
○ *I'll count by 5, 13 times to get the product.*

Arrays

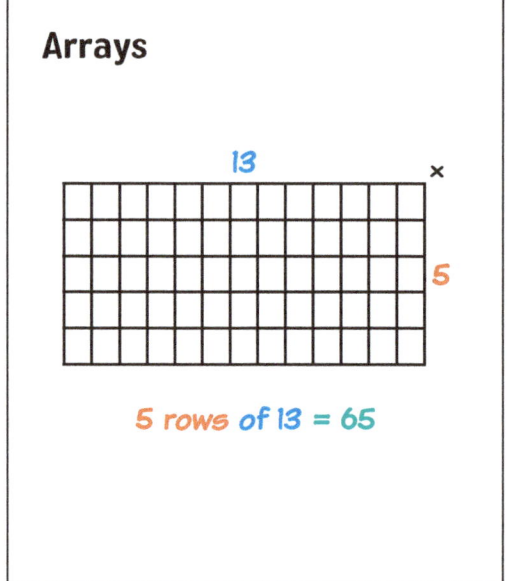

5 rows of 13 = 65

Decomposition/ Distributive Property

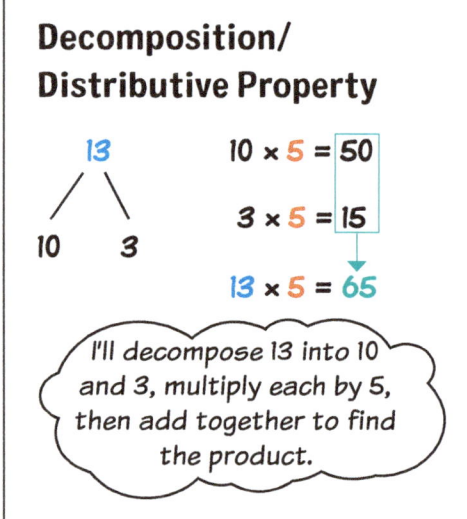

I'll decompose 13 into 10 and 3, multiply each by 5, then add together to find the product.

Area Model

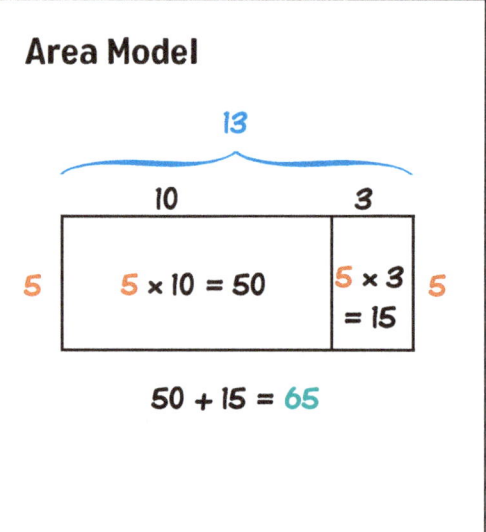

50 + 15 = 65

Partial Product

```
 13    = 10 + 3
× 5
 15    = 3 × 5
+50    = 10 × 5
 65
```

- Decompose 13 into 1 ten plus 3 ones.
- Multiply each part of the decomposed number by 5.
- Add together each partial product.
- Now you have the product (answer).

Multi-Digit Multiplication Strategies (continued)

Decomposition/Distributive Property

21 × 34 = ☐

20 + 1   30 + 4

(20 × 30) + (20 × 4) + (1 × 30) + (1 × 4)

600 + 80 + 30 + 4

680 + 34

714

- Decompose (break apart) each number into tens and ones.
- Multiply each part of the first number by each part of the second number.
- Add together each partial product.
- Now you have the product (answer).

💡 Working with even bigger numbers!

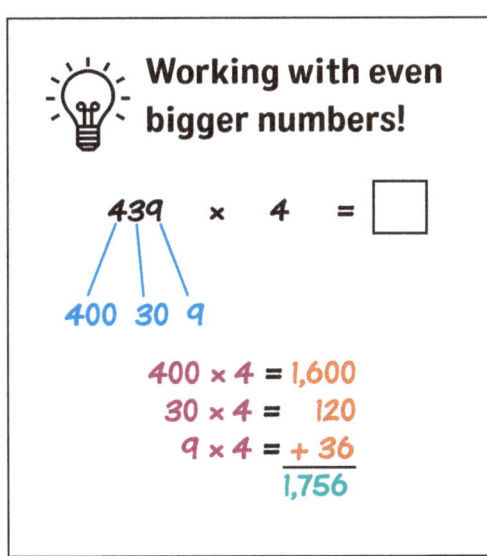

Partial Product

I use this same method to solve using partial product!

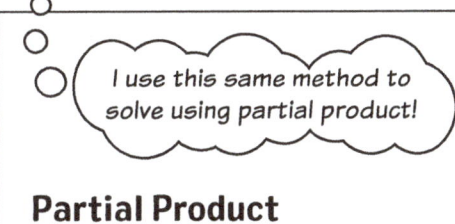

$$
\begin{array}{r}
21 = (20+1) \\
\times\,34 = (30+4) \\
\hline
4 = 4 \times 1 \\
80 = 4 \times 20 \\
30 = 30 \times 1 \\
+\,600 = 30 \times 20 \\
\hline
714
\end{array}
$$

Area Model

81 × 615 = ☐

	600	10	5
80	80 × 600 = 48,000	80 × 10 = 800	80 × 5 = 400
1	1 × 600 = 600	1 × 10 = 10	1 × 5 = 5

$$
\begin{array}{r}
^{1} \\
48{,}000 \\
800 \\
400 \\
600 \\
10 \\
+\quad 5 \\
\hline
49{,}815
\end{array}
$$

Multiplication and Division Using Place Value

Video 8: Multi-Digit Multiplication With Base-Ten Pieces: You can use pre-grouped and groupable base-ten pieces to help students connect multiplication strategies and algorithms. *Pictured here: base-ten blocks and KP® Ten-Frame Tiles*

https://qrs.ly/tgg99kt

Understanding the Traditional Multiplication Method

To use the **traditional multiplication method** (or standard U.S. algorithm), multiply from right to left, regrouping as necessary.

```
  24
x 34
-----
   ?
```

Word Problem

There are **34** students in the Nature Club who want to take a field trip to the botanical gardens and wildlife preserve. It costs **$24** per student for admission. How much would admission cost for all **34** students?

Step 1

```
  1
  24
x 34
-----
   6
```

4 ones × **4** ones = **16** (**1** ten and **6** ones). Record **6** beneath the ones column and regroup **1** ten.

Step 2

```
  1
  24
x 34
-----
  96
```

4 ones × **2** tens = **8** tens. Add the **1** ten that was regrouped and record the **9** tens.

Step 3

```
 1 1
  24
x 34
-----
  96
  20
```

3 tens × **4** ones = **12** tens (**1** hundred and **2** tens, or **20**). Record the **20** beneath the first product and regroup the **1** hundred.

Step 4

```
 1 1
  24
x 34
-----
  96
 720
```

3 tens × **2** tens = **6** hundreds Add the **1** hundred that was regrouped and record the **7** hundreds.

Step 5

```
 1 1
  24
x 34
-----
  96
+720
-----
 816
```

6 ones + **0** ones = **6** ones
9 tens + **2** tens = **11** tens (**1** hundred and **1** ten) **7** hundreds + the **1** hundred that was regrouped = **8** hundreds

---

Traditional Method

```
 1 1
  24
x 34
-----
  96
+720
-----
 816
```

Multiplying Decimals

You **multiply numbers with digits on both sides of the decimal point** the same way that you multiply whole numbers. The only thing that differs is determining where you place the decimal point in your product.

You can use an area model to multiply decimals.

2.4 x 3.8 = ☐

Represent **2.4** as **2 wholes** and **4 tenths** horizontally and **3.8** as **3 wholes** and **8 tenths** vertically on the same area model.

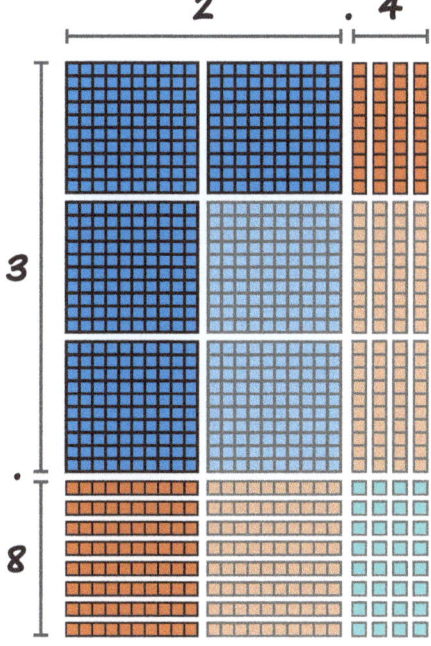

Fill in the area to complete the rest of the rectangle. Add up the number of units from each action to calcuate the product.

6 wholes + 12 tenths + 16 tenths + 32 hundredths

6 + 1.2 + 1.6 + 0.32

9.12

You can also use what you know about multiplication of whole numbers and estimation to solve multiplication problems with decimals.

Multiplication and Division Using Place Value

Multiplying Decimals (continued)

1.5 × 3.49 = ☐

Make an Estimate

1.5 is close to **2**, while **3.49** is close to **3**. My product is about **2 × 3**, which is **6**.

Multiply

Multiply as if these numbers are whole numbers*:

```
    2 4
    3.49
  ×  1.5
  1 1
   1,745
 + 3,490
   5,235
```

*See page 66.

Place the Decimal Point

The estimate was **6**, so it would make sense to place the decimal point after the **5**:

5.235

 Look! A helpful mathematical discovery!

Place the decimal point in the product by counting the number of places to the right of the decimal in each factor. The total tells you the number of places there will be to the right of the decimal in your product!

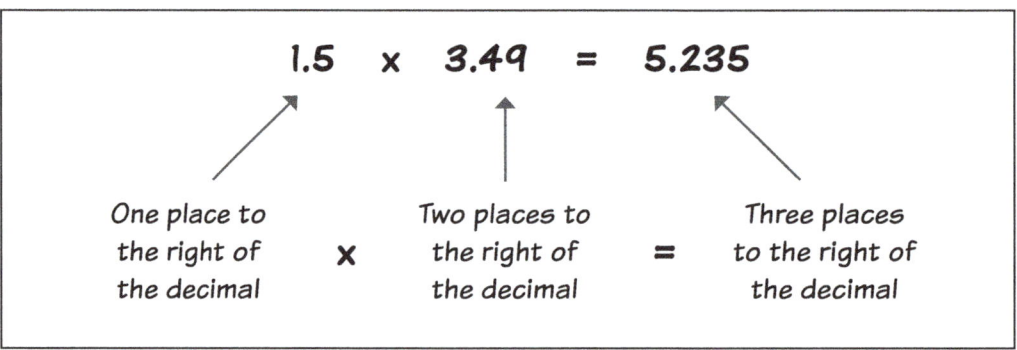

1.5 × 3.49 = 5.235

One place to the right of the decimal × Two places to the right of the decimal = Three places to the right of the decimal

0.3 × 0.4 = 0.12

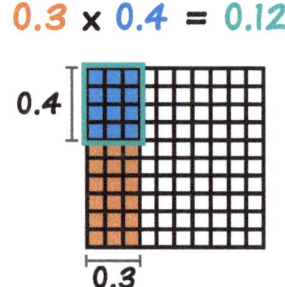

When I multiply a fractional part (0.3) by another fractional part (0.4), I have a part of a part of the whole (0.12), which is less than what I started with!

Division Word Problems..................

Division is the opposite of multiplication and you use multiplication every time you divide. When you divide, you are determining the number of equal shares or size of each share. There are four division problem types:

Division as Sharing

If 12 toys are divided equally among 3 brothers, how many toys will each brother receive?

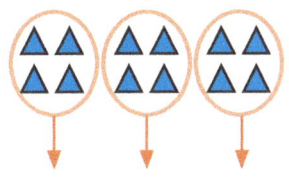

$12 \div 3 = 4$

Arrays of Objects

If 12 apples are arranged into an array with 3 rows, how many apples will be in each row?

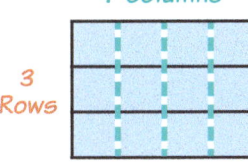

$12 \div 3 = \square$

$12 \div 3 = 4$

Compare

An orange hat costs $12, and a blue hat costs $3. How many times more expensive is the orange hat than the blue hat?

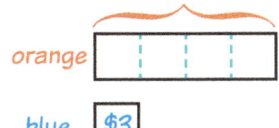

$12 \div \square = 3$

$12 \div 4 = 3$

Division as Grouping

12 toys are given equally to some friends. Each friend receives 3 toys.

How many friends received the toys?

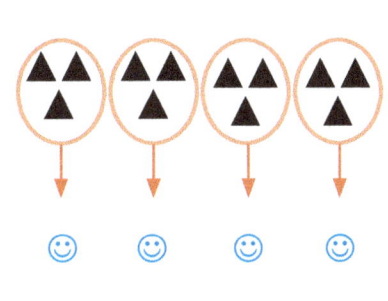

$12 \div 3 = 4$

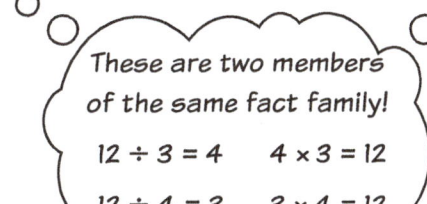

These are two members of the same fact family!

$12 \div 3 = 4 \quad 4 \times 3 = 12$

$12 \div 4 = 3 \quad 3 \times 4 = 12$

Multiplication and Division Using Place Value

Fractions as Division

Fractions are another way to represent a division situation. When two whole numbers are divided, the answer can be a whole number, a mixed number, or a fraction. This example shows a **whole number** answer.

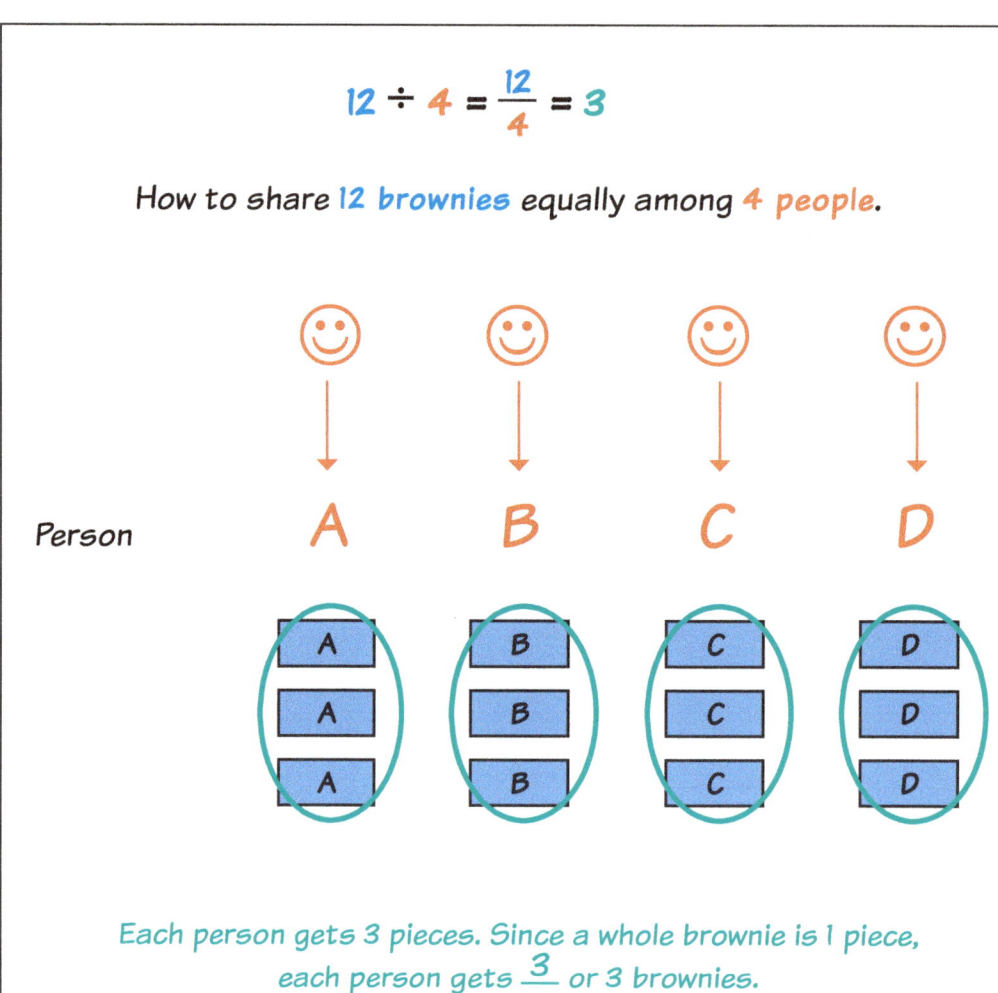

Fractions as Division (continued)

When two whole numbers are divided, the answer can be a **mixed number**.

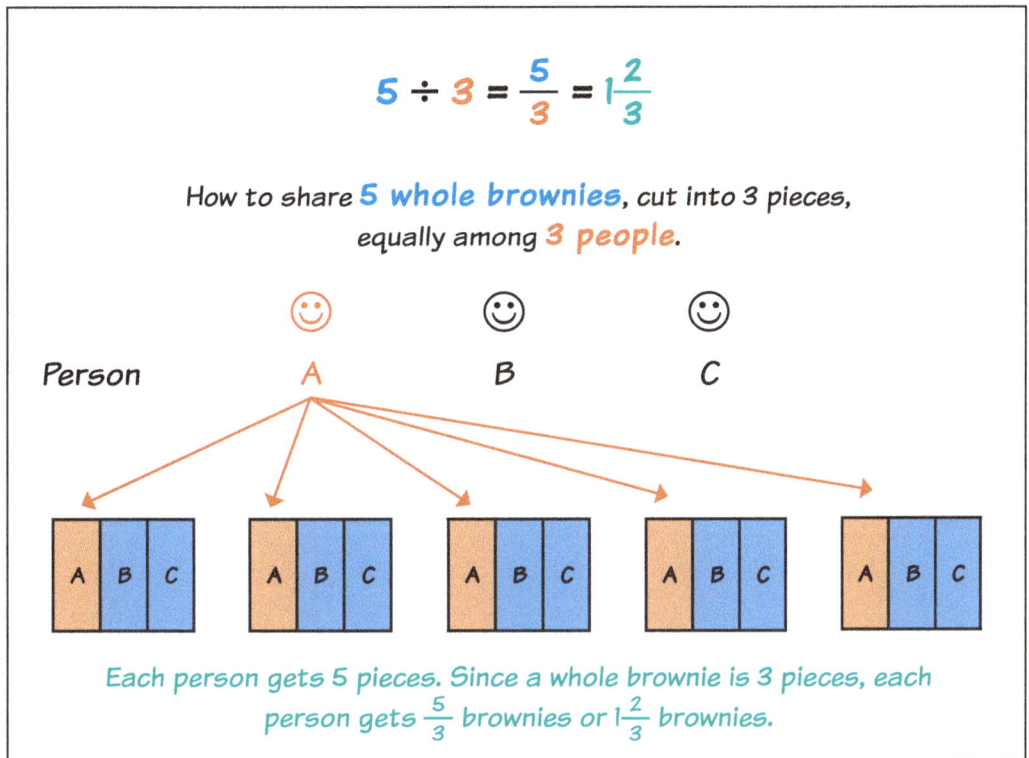

When two whole numbers are divided, the answer can be a **fraction**.

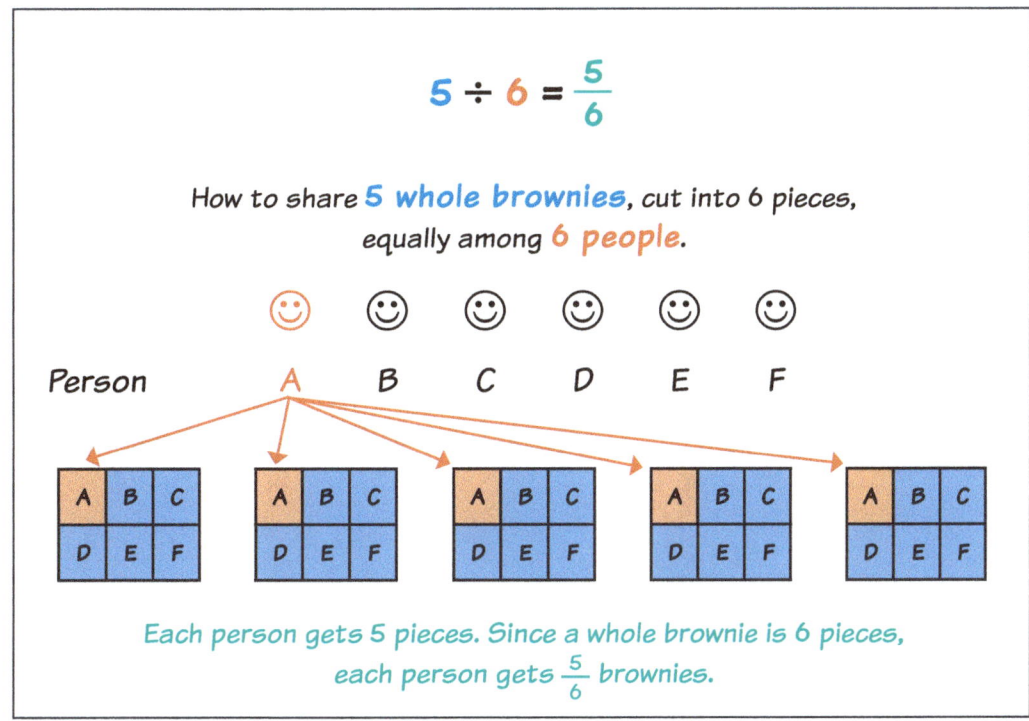

Multiplication and Division Using Place Value

Division Strategies

There are a few different **division strategies**. Three common strategies are detailed below:

$$966 \div 7 = \boxed{}$$

Area Model : Finding Side Length

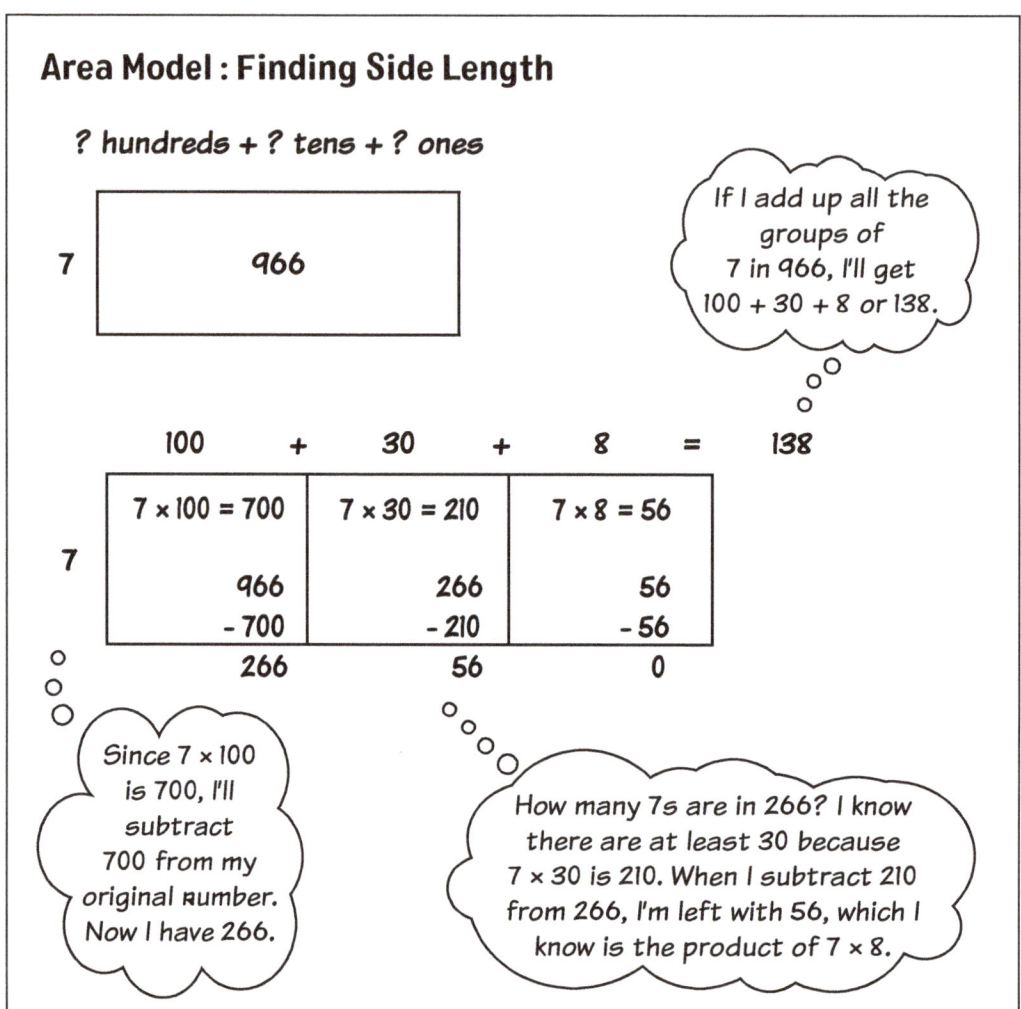

Partial Quotient Method

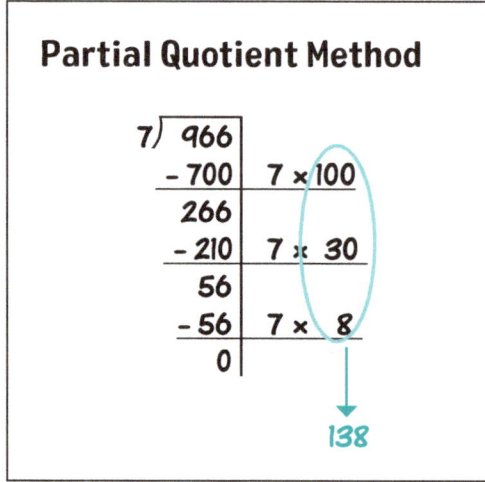

Traditional Method

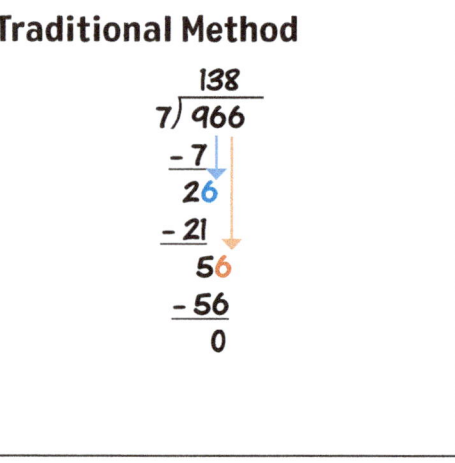

Understanding the Partial Quotient Division Method

The **partial quotient** division method uses repeated subtraction of friendly factors to find partial answers to the problem. Once you have reached zero, these partial products are added together to find the final answer.

Video 9: Multi-Digit Division With Base-Ten Pieces: You can use pre-grouped and groupable base-ten pieces to help students connect division strategies and algorithms. *Pictured here: base-ten blocks and KP® Ten-Frame Tiles*

https://qrs.ly/utg99kv

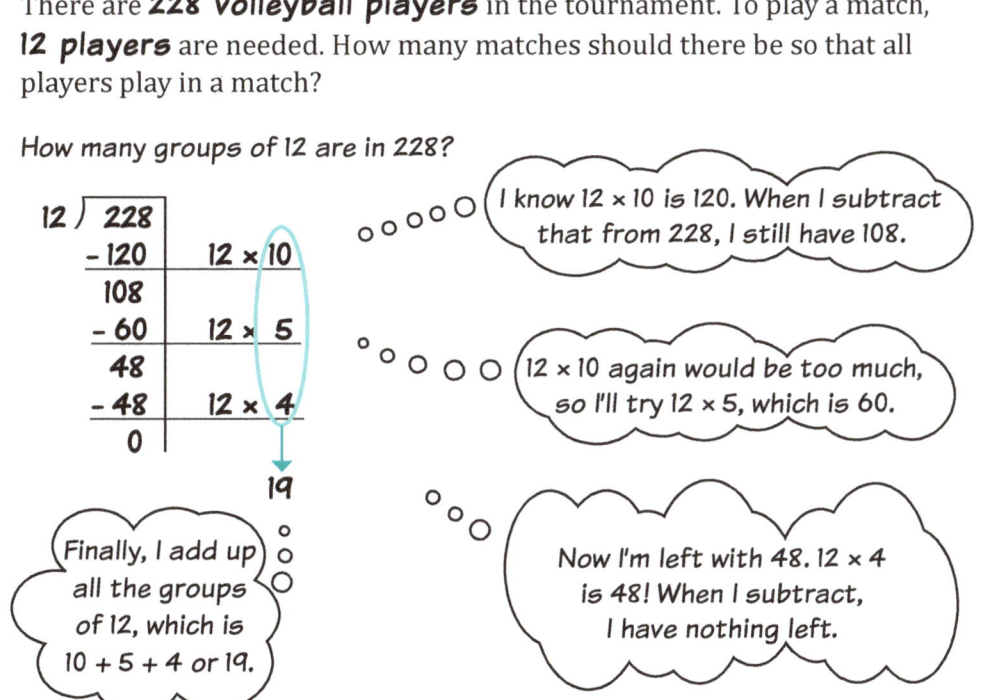

There are **228 volleyball players** in the tournament. To play a match, **12 players** are needed. How many matches should there be so that all players play in a match?

How many groups of 12 are in 228?

```
12 ) 228
   - 120    12 × 10
     108
    - 60    12 × 5
      48
    - 48    12 × 4
       0
              19
```

I know 12 × 10 is 120. When I subtract that from 228, I still have 108.

12 × 10 again would be too much, so I'll try 12 × 5, which is 60.

Now I'm left with 48. 12 × 4 is 48! When I subtract, I have nothing left.

Finally, I add up all the groups of 12, which is 10 + 5 + 4 or 19.

Partial Quotient Method

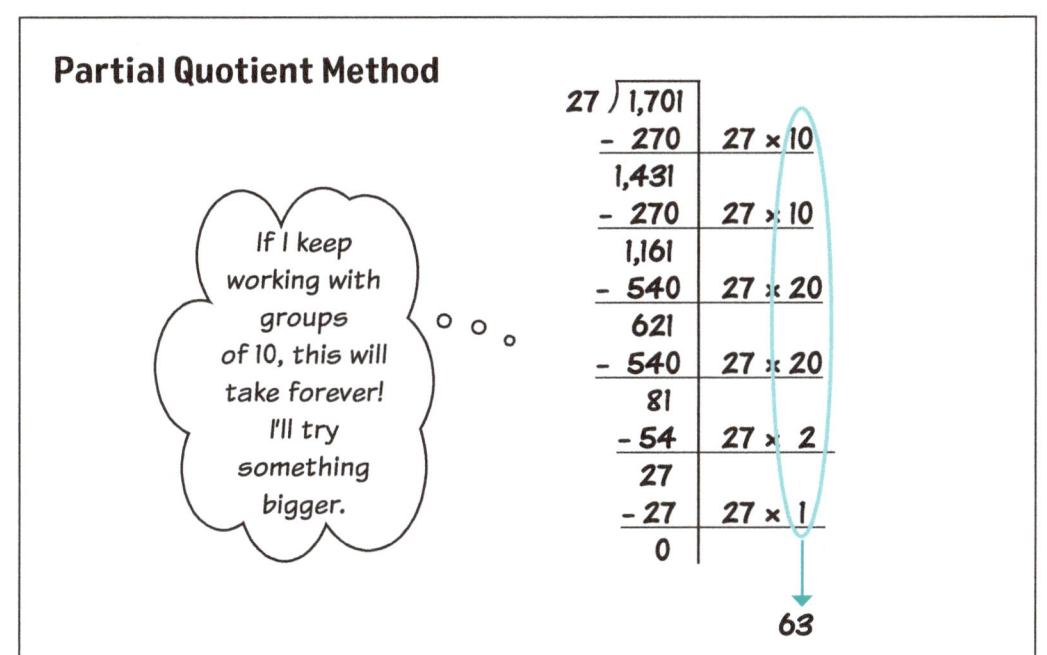

```
27 ) 1,701
    -  270    27 × 10
     1,431
    -  270    27 × 10
     1,161
    -  540    27 × 20
       621
    -  540    27 × 20
        81
    -   54    27 × 2
        27
    -   27    27 × 1
         0
                63
```

If I keep working with groups of 10, this will take forever! I'll try something bigger.

Multiplication and Division Using Place Value

Understanding the Traditional Division Method

The traditional U.S. division method is often called **long division**. Long division is a method for dividing that repeats the basic steps: ① Divide; ② Multiply; ③ Subtract; ④ Drop down the next digit. In long division, you work from left to right.

You've got 520 players. It takes 8 players to form a tug-of-war team.

How many teams can you make?

```
    _65
8)  520
   -48↓
     40
    -40
      0
```

- *Start with the hundreds.*
  - *8 × 100 = 800*
  - *The dividend is only 500, so nothing will go "above" the 5 hundreds.*

- *Divide the tens.*
  - *8 × 6 tens = 48 tens*
  - *8 × 7 tens = 56 tens*
  - *So, the quotient is between 60 and 70. Write a 6 in the tens place and 48 (tens) under the dividend. Subtract.*

- *Divide the ones.*
  - *8 × 5 = 40*
  - *Write a 5 in the ones place and 40 under the dividend. Subtract.*

Traditional Method

```
       19
  12) 228
     -12↓
      108
     -108
        0
```

Traditional Method

```
        63
  27) 1,701
      -162↓
         81
        -81
          0
```

Remainders

Sometimes when you are dividing you will discover that your **divisor** does not go evenly into your **dividend**. In this case, you may have extras or **remainders** left to deal with.

This is my remainder.

In order to determine what to do with the **remainder**, you must consider the context of the problem.

Round-Up Remainder

If there are **129 students** and **4** fit into each car, how many cars are needed?

4 × 32 = 128

Remainder of 1 means 1 is left behind! You will therefore need 33 cars.

Fraction Remainder

If there are **129 cookies** to be shared equally among **4 classes**, how many cookies would each class receive?

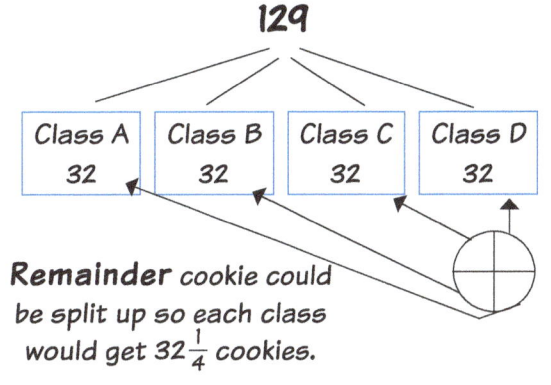

Remainder cookie could be split up so each class would get $32\frac{1}{4}$ cookies.

Whole Number Remainder

If there are **129 balloons** and it takes **4 balloons** to make a balloon bouquet, how many full bouquets can be made?

4 ballons × 32 bouquets = 128 balloons
1 ballon left over
You can make 32 FULL bouquets

Decimal Remainder

```
   0.3225
4 )1.2900
  -12
   09
   - 8
    10
   - 8
    20
   -20
     0
```

If bananas cost **$1.29** for **4 bananas** and I purchase 1 banana, how much will it cost?

The store wouldn't want to lose the extra **remainder** of a penny, so it would cost **$0.33** for one banana.

Source: Balloons image from iStock.com/Photoplotnikov

Multiplication and Division Using Place Value

Dividing Decimals

You **divide numbers with digits on both sides of the decimal point** the same way you divide whole numbers. You only need to determine where to place the decimal point in your quotient.

Decimal ÷ Whole Number

```
        6.72
   4 ) 26.88
      -24
        2 8
       -2 8
          08
         - 8
           0
```

Place the decimal point directly above the decimal in the dividend.

Decimal ÷ Decimal or Whole Number ÷ Decimal

You can change your division problem into one with only whole numbers by multiplying the divisor and dividend by a power of ten.

dividend	divisor	quotient
24	0.16	150
240	1.6	150
2,400	16	150

$$0.16 \overline{\smash{)}24} = \frac{24}{0.16} \times \frac{100}{100} = \frac{2400}{16} = 150$$

Then divide the same way you would divide whole numbers!

Written traditionally, it would look like this:

$$0.16 \overline{\smash{)}24.00}$$

Shift the digits two places to the left to show 0.16 × 100.

Shift the digits two places to the left to show 24 × 100.

Then, divide as usual.

```
           150
     16 ) 2400
         -16
           80
          -80
           00
          - 0
            0
```

The solution to $0.16 \overline{\smash{)}24}$ is the same as the solution to $16 \overline{\smash{)}2400}$, which is 150.

Two-Step Word Problems (+ − × ÷)

Two-step word problems can require any two of the basic operations (+, −, ×, ÷) to solve the problem. You can use the **order of operations** (see page 27) to know which operation to do first.

Sam is 2 years older than 3 times Katie's age. Katie is 5 years old. How old is Sam?

Katie | 5

Sam | 5 | 5 | 5 | 2

? (17)

$(3 \times 5) + 2 = \square$

Step 1: $(3 \times 5) + 2 = \square$

Step 2: $15 + 2 = \boxed{17}$

Sam is 17 years old.

---

331 students went on a field trip. 6 buses were filled and 7 students went in cars. How many students were on each bus?

| 331 |
| 54 | 54 | 54 | 54 | 54 | 54 | 7 |

? (54)

$\dfrac{331 - 7}{6} = \square$

Step 1: $\dfrac{331 - 7}{6} = \square$

Step 2: $324 \div 6 = \boxed{54}$

There were 54 students on each bus.

Alyiah had $24 to spend on 7 identical pens. After buying them, she still had $10. How much did each pen cost?

| 24 |
| 2 | 2 | 2 | 2 | 2 | 2 | 2 | 10 |

?

$(\$24 - \$10) \div 7 = \square$

Step 1: $(\$24 - \$10) \div 7 = \square$

Step 2: $\$14 \div 7 = \boxed{\$2}$

Each pen cost $2.

Kim bought a magazine for $5 and 4 balloons. She spent a total of $25. How much did each balloon cost?

?

| 5 | 5 | 5 | 5 | 5 |
| 25 |

$\$5 + (4 \cdot \square) = \$25$

Step 1:
$\$5 - \$5 + (4 \cdot \square) = \$25 - \$5$
$0 + (4 \cdot \square) = \$20$

Step 2:
$4 \cdot \square \div 4 = \$20 \div 4$
$\square = \$5$

Each balloon cost $5.

 Whatever you do to one side of an equation, you must also do to the other side to keep the equation balanced!

Multiplication and Division Using Place Value

Chapter 6

Special Topics With Whole Numbers

Math is a study of patterns. And the study of math encourages a focus on identifying and studying recurring regularities in all sets of numbers: whole numbers, decimals, fractions, and so on.

Many whole numbers have special traits, or properties, that help children identify patterns. For example, numbers can be even or odd, prime or composite. They can even form a perfect square when thinking about an area model for multiplication.

When students recognize these special traits, they are empowered to use them to solve problems with whole numbers. It's important to understand that the more children recognize and talk about these special traits, the more they'll be able to use them to solve increasingly complex problems.

Typical Trajectory in Most State Standards Frameworks:

- Grade 2: Introduce number properties such as even and odd
- Grade 3: Introduce number properties such as perfect squares (using an area model)
- Grades 4–5: Introduce number properties such as factors, multiples, prime, and composite
- Grades 5–6: Introduce exponents as powers of ten
- Grade 6: Introduce exponents as powers of any number

Odd and Even Numbers

Video 10: Even and Odd With Ten Frames and Counters: You can use different models to help students "see" the concept of even and odd. *Pictured here: KP Ten-Frame Tiles virtual app*

https://qrs.ly/jug99l2

https://qrs.ly/y7g99jp

An **even number** is any number that can make groups of 2 with no leftover units or can be made into an array 2 units high.

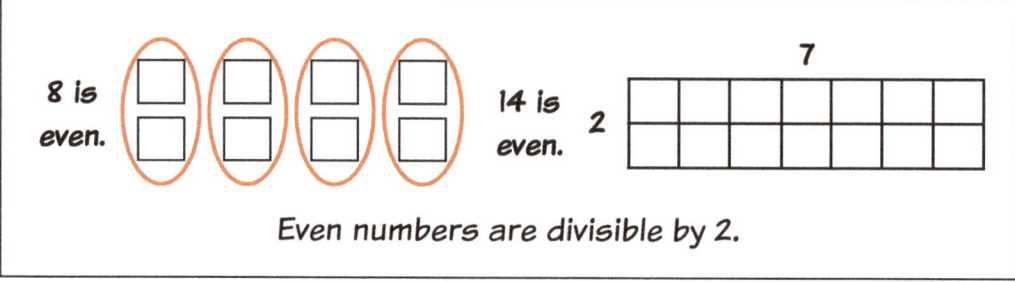

Even numbers are divisible by 2.

An **odd number** is any number that can not be made into an array 2 units high without a leftover unit. An odd number can be split up into partners or pairs, but will always have a leftover unit.

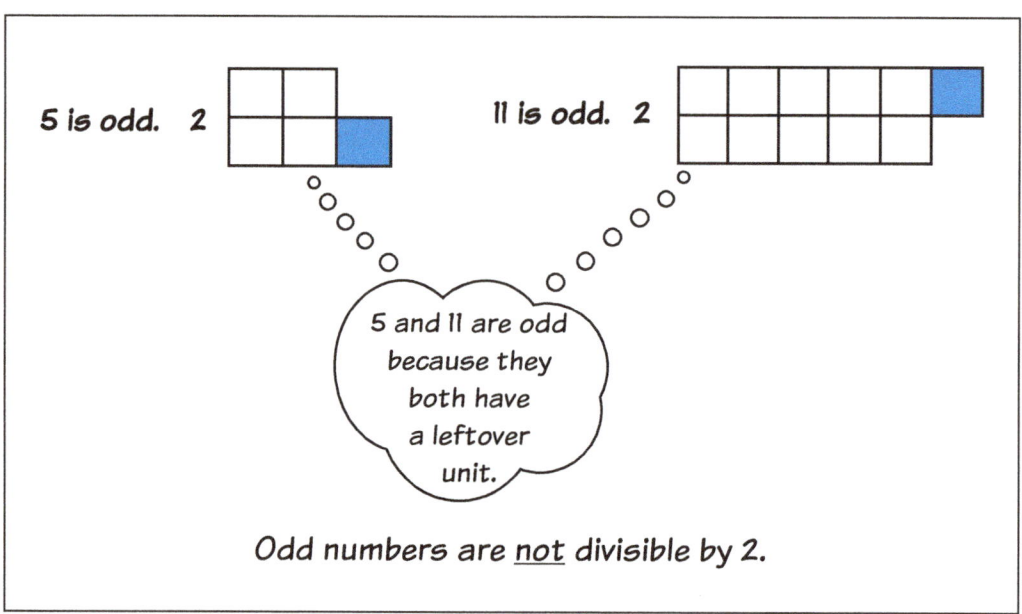

5 and 11 are odd because they both have a leftover unit.

Odd numbers are <u>not</u> divisible by 2.

Even numbers end in a
2, 4, 6, 8, or 0.

Odd numbers end in a
1, 3, 5, 7, or 9.

Hundreds Chart

I can use a **hundreds chart** to explore lots of different math ideas and see number patterns.

1	2	3	4	5	6	7	8	9	10
11	12	13	14	15	16	17	18	19	20
21	22	23	24	25	26	27	28	29	30
31	32	33	34	35	36	37	38	39	40
41	42	43	44	45	46	47	48	49	50
51	52	53	54	55	56	57	58	59	60
61	62	63	64	65	66	67	68	69	70
71	72	73	74	75	76	77	78	79	80
81	82	83	84	85	86	87	88	89	90
91	92	93	94	95	96	97	98	99	100

> An example is seeing that all of the even numbers are shaded, while the odds are not.
>
> Another example is noticing that the multiples of 3, which are circled in orange, alternate between being even and odd numbers.

This drawing shows the seats in a theater. Sam noticed a pattern in the seat numbers. The number of each seat is 10 more than the one below it. What number would be on the seat that is marked with an **X**?

			X						
41	42	43	44	45	46	47	48	49	50
31	32	33	34	35	36	37	38	39	40

> 10 more than 44 is 54, so the seat number must be **54**!

Special Topics With Whole Numbers

Multiples and Common Multiples

A **multiple** is a number included when you skip-count by a certain amount. It is the result of multiplying a whole number by an integer (see page 90 for more about integers).

Positive Multiples of 12

$0 = 12 \times 0$
$12 = 12 \times 1$
$24 = 12 \times 2$
$36 = 12 \times 3$
$48 = 12 \times 4$
$60 = 12 \times 5$
$72 = 12 \times 6$
etc.

*The positive **multiples** of 12 are: 0, 12, 24, 36, 48, 60, 72,... because they all are products of 12 times an integer.*

 Notice that 6 is not a multiple of 12 because you cannot multiply 12 by a whole number to get a product of 6.

×	1	2	3	4	5	6	7	8	9	10
1	1	2	3	4	5	6	7	8	9	10
2	2	4	6	8	10	12	14	16	18	20
3	3	6	9	12	15	18	21	24	27	30
4	4	8	12	16	20	24	28	32	36	40
5	5	10	15	20	25	30	35	40	45	50
6	6	12	18	24	30	36	42	48	54	60
7	7	14	21	28	35	42	49	56	63	70
8	8	16	24	32	40	48	56	64	72	80
9	9	18	27	36	45	54	63	72	81	90
10	10	20	30	40	50	60	70	80	90	100

On this multiplication table, all of the multiples of 4 have been shaded in pink. If you skip count by 4 you will say each of these numbers.

When comparing numbers, it is often necessary to identify their **common multiples**.

Find the **common multiples** for 12 and 15.

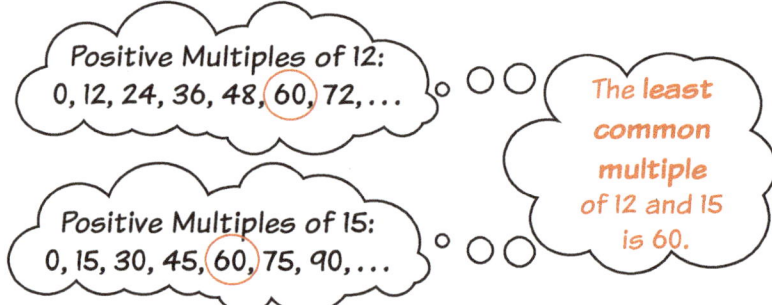

The smallest multiple (other than zero) that two numbers share is called their **least common multiple**.

Factors

Factors are the whole numbers that are multiplied together to get a product. A number is divisible by each of its factors.

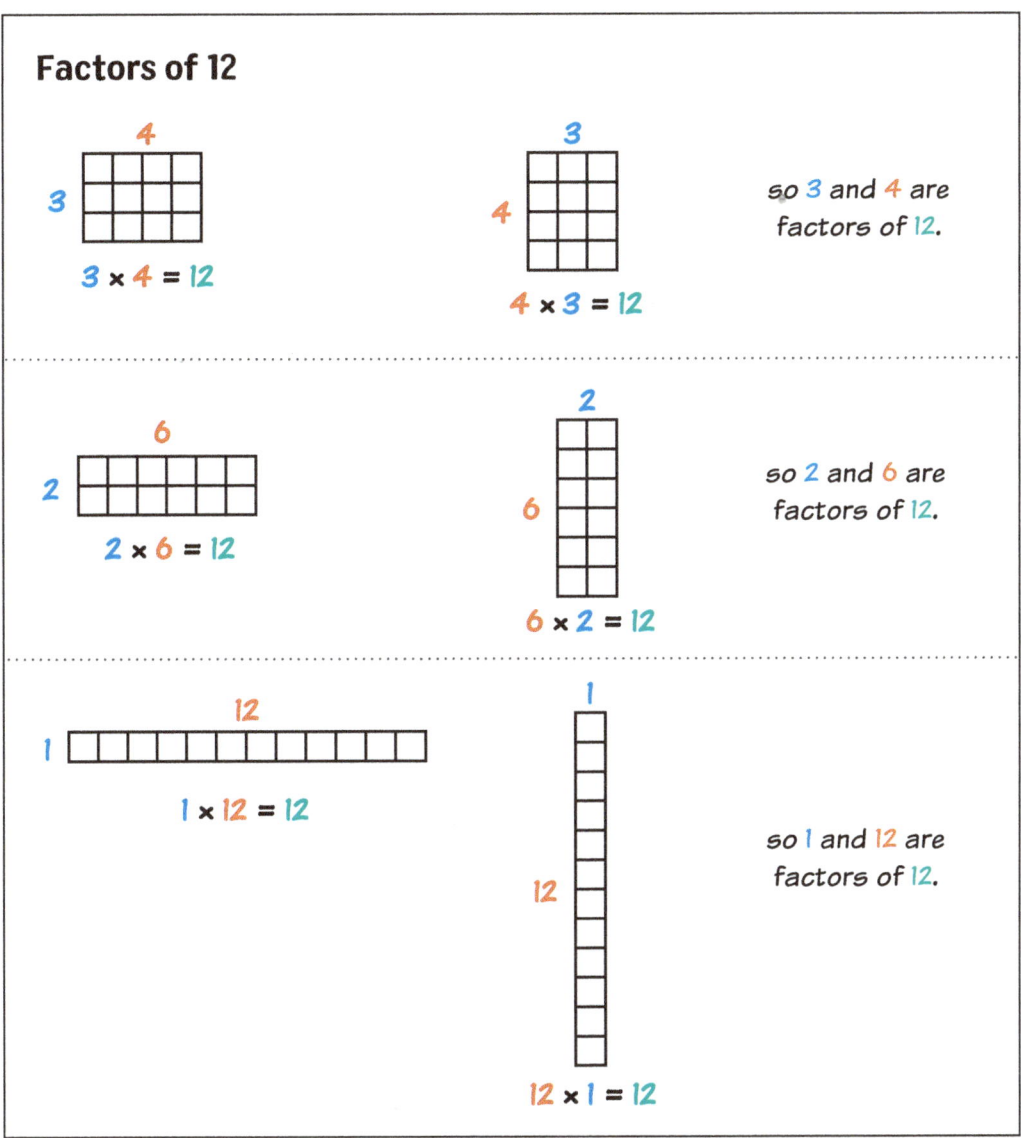

The **factors** of 12 are 1, 2, 3, 4, 6, and 12 because they all go into 12 evenly.

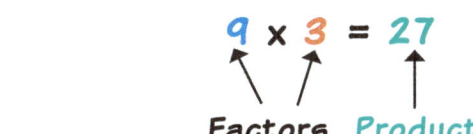

Factors   Product

The answer to a multiplication problem is called a *product*.

Special Topics With Whole Numbers

Common Factors

When comparing numbers, it is often necessary to identify their **common factors**.

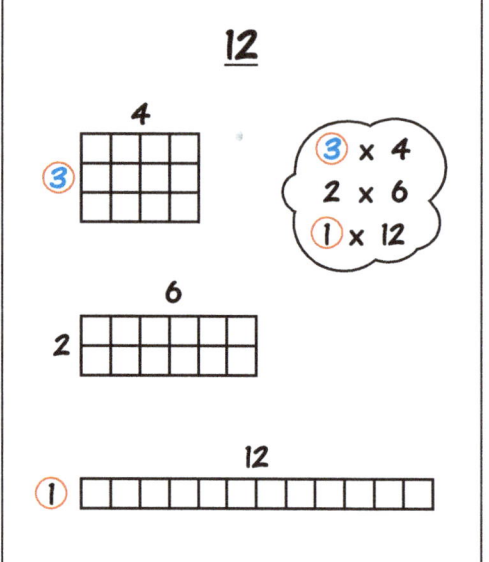
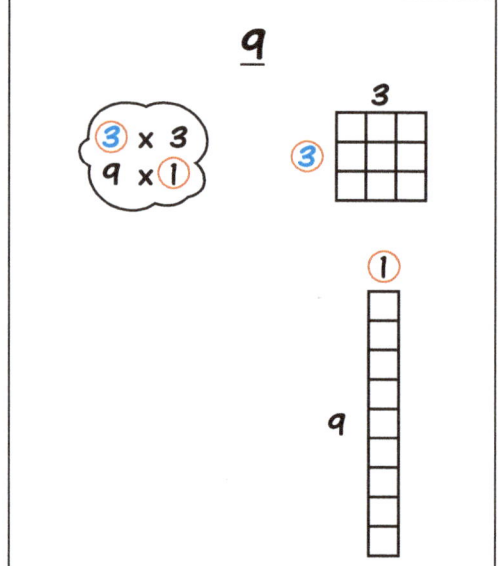

The **common factors** of 12 and 9 are 1 and 3, since these are the factors they share.

The **greatest common factor** of 12 and 9 is 3, since 3 is the largest of their shared factors.

Prime Numbers

Almost all natural numbers (numbers you use when you count) can be described as **prime numbers** or **composite numbers**.

Prime numbers have exactly two factors: 1 and the number itself.

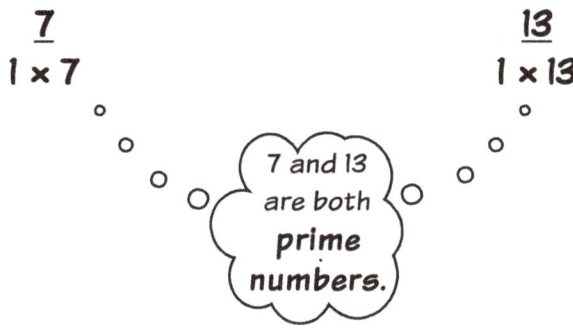

7
1 × 7

13
1 × 13

7 and 13 are both prime numbers.

Prime numbers have exactly two factors, so they can make only two arrays.

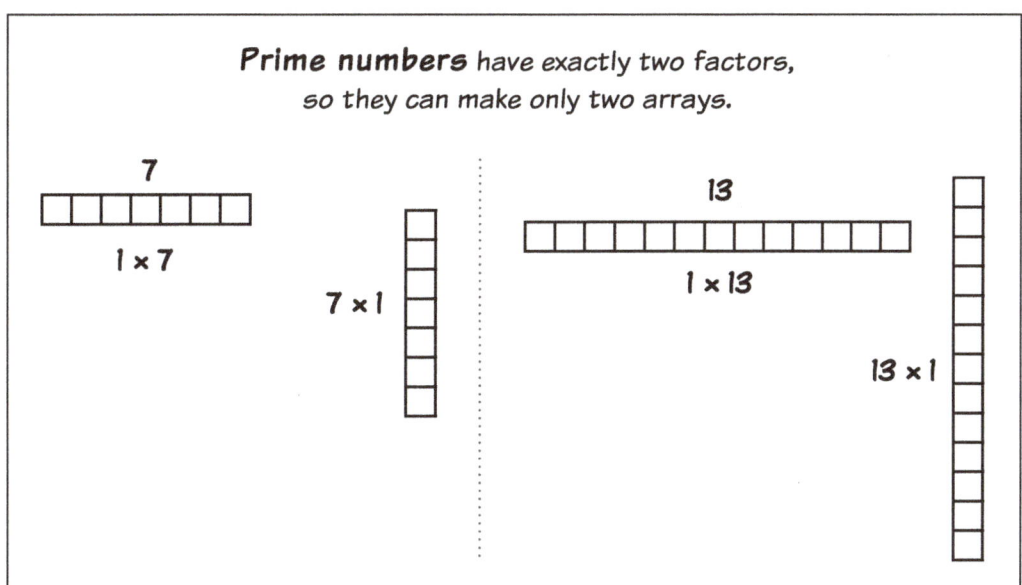

7
1 × 7

7 × 1

13
1 × 13

13 × 1

The **prime numbers** from 1 to 50 are circled below:

1, ②, ③, 4, ⑤, 6, ⑦, 8, 9, 10, ⑪, 12, ⑬, 14, 15, 16, ⑰, 18, ⑲, 20, 21, 22, ㉓, 24, 25, 26, 27, 28, ㉙, 30, ㉛, 32, 33, 34, 35, 36, ㊲, 38, 39, 40, ㊶, 42, ㊸, 44, 45, 46, ㊼, 48, 49, 50

Special Topics With Whole Numbers

Composite Numbers

Most whole numbers are not **prime numbers**. Instead, they are **composite numbers**.

Composite numbers have more than two factors.

<u>16</u>
1 × 16
2 × 8
4 × 4

16 and 15 are examples of composite numbers.

<u>15</u>
1 × 15
3 × 5

Composite numbers have three or more factors; therefore, they can make more than two arrays.

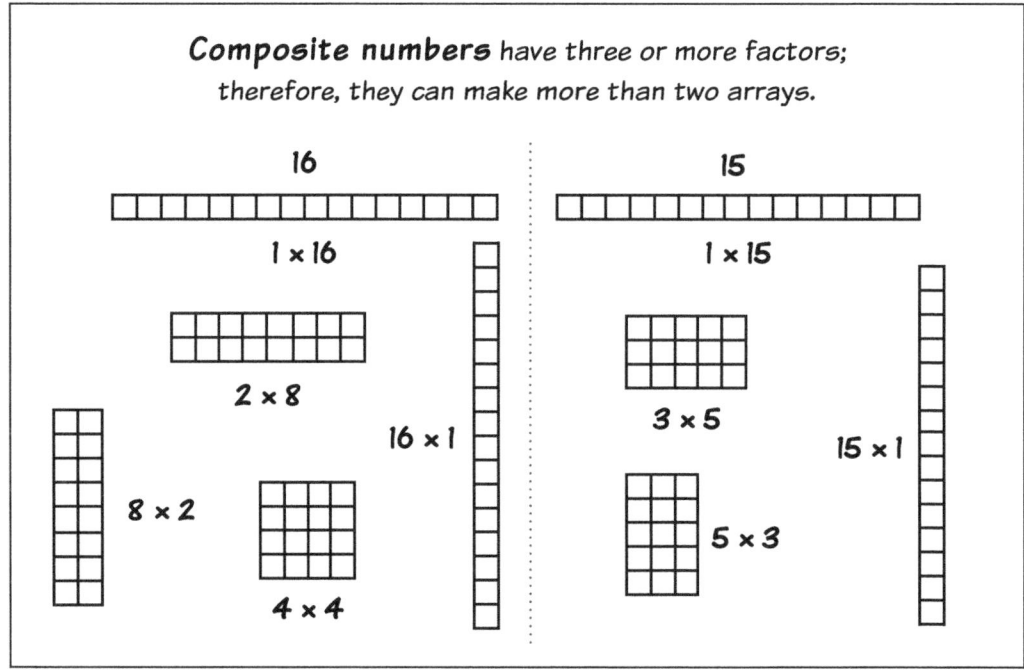

---

1 Is a Special Case

1 is neither **prime** nor **composite** because it only has one unique factor and can only make one array.

<u>1</u>
1 × 1

1 × 1

Exponents................................

Exponents (powers) are a short way to write repeated multiplication of the same number.

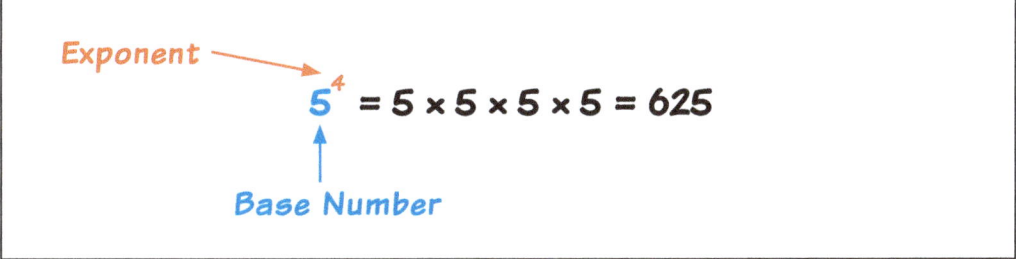

Exponent

$$5^4 = 5 \times 5 \times 5 \times 5 = 625$$

Base Number

👉 The exponent tells you how many times the base number is being multiplied by itself.

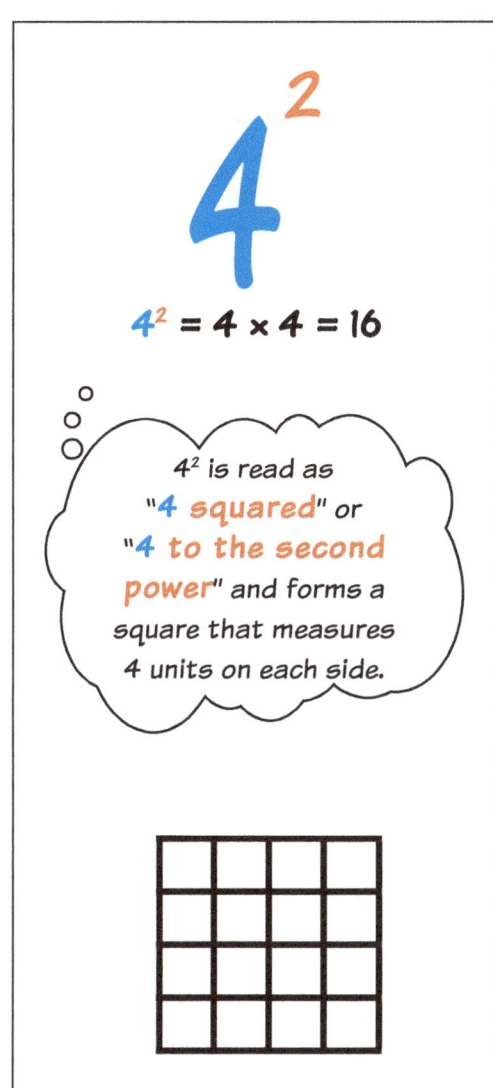

4^2

$4^2 = 4 \times 4 = 16$

4^2 is read as "**4 squared**" or "**4 to the second power**" and forms a square that measures 4 units on each side.

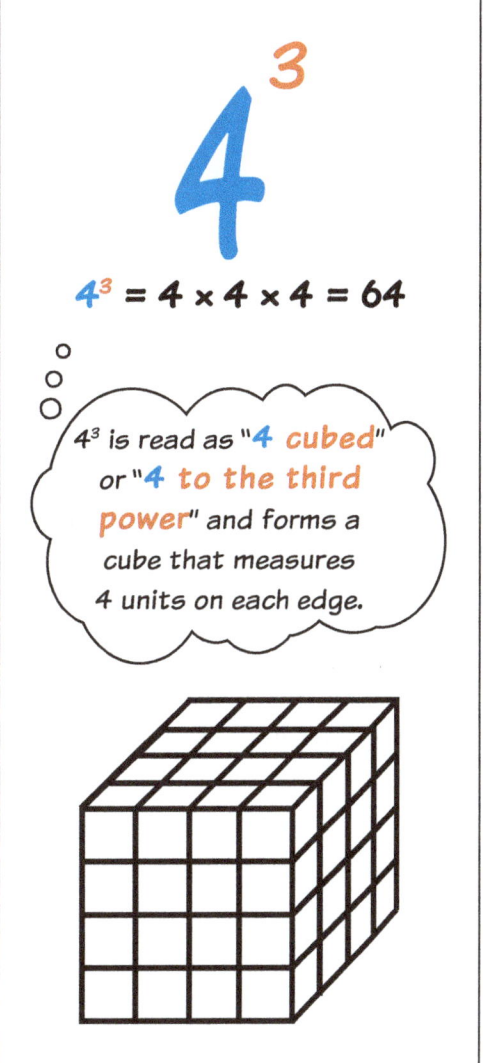

4^3

$4^3 = 4 \times 4 \times 4 = 64$

4^3 is read as "**4 cubed**" or "**4 to the third power**" and forms a cube that measures 4 units on each edge.

Special Topics With Whole Numbers

Chapter 7

Negative and Positive Numbers

Negative and positive numbers, including integers, are seen in daily life and in math. Students sometimes struggle with the idea that a number can be less than zero, so sharing examples like the following, from everyday life, can help them to understand this idea:

- When the summer's daily temperature reaches 110°, that is a positive number.

- When we purchase a book for $9.95, our bank account records a negative number.

- When the score in a video game goes up, it is a positive number, like 40,000; however, when a player misses, the score is decreased, and the number of points decreased are shown as a negative number, like −10,000.

All numbers, except for zero, are either negative or positive. That said, not all numbers are integers. Integers include all whole numbers and their opposites. Think of a number line with all whole numbers to the right of the decimal point. If you fold the number line in half at zero, then the whole numbers will match up to their opposites on the negative side of the number line. This is the set of all integers, both positive and negative.

Numbers such as fractions and decimals can also be positive or negative, but they are not integers unless they are equivalent to a whole number or its opposite.

Typical Trajectory in Most State Standards Frameworks:

- Grades K–5: All positive whole numbers are gradually introduced, though they are never called "integers"
- Grade 6: All whole numbers are called "integers" for the first time, and students are introduced to all negative integers

Integers

Integers are a set of whole numbers and their opposites.

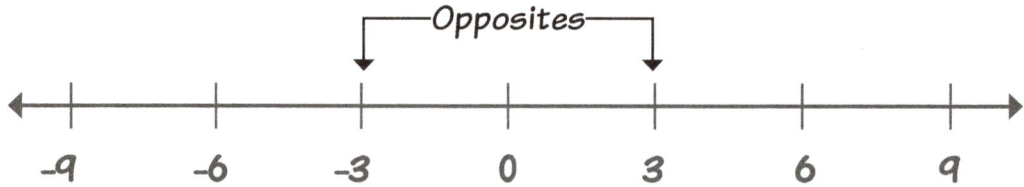

- −3 and 3 are opposites.
- Zero is its own opposite.

Positive and Negative Numbers

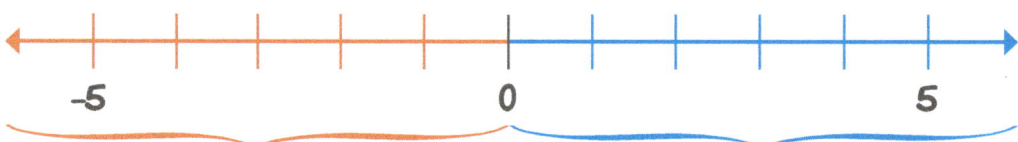

Negative Numbers Are Less Than Zero

Examples of Negative Numbers

$-9, -4\frac{1}{5}, -2.63$

−5

- Molly owes her sister $5.

−2

- Seth's parents were married 2 years before he was born.

Positive Numbers Are Greater Than Zero

Examples of Positive Numbers

$12, \pi, \$9.46$

+23

- Jane earned $23 babysitting last weekend.

+10,000

- The airplane reached its cruising altitude of 10,000 feet.

Comparing Integers

You can use a **number line** to compare integers.

Horizontal Number Line

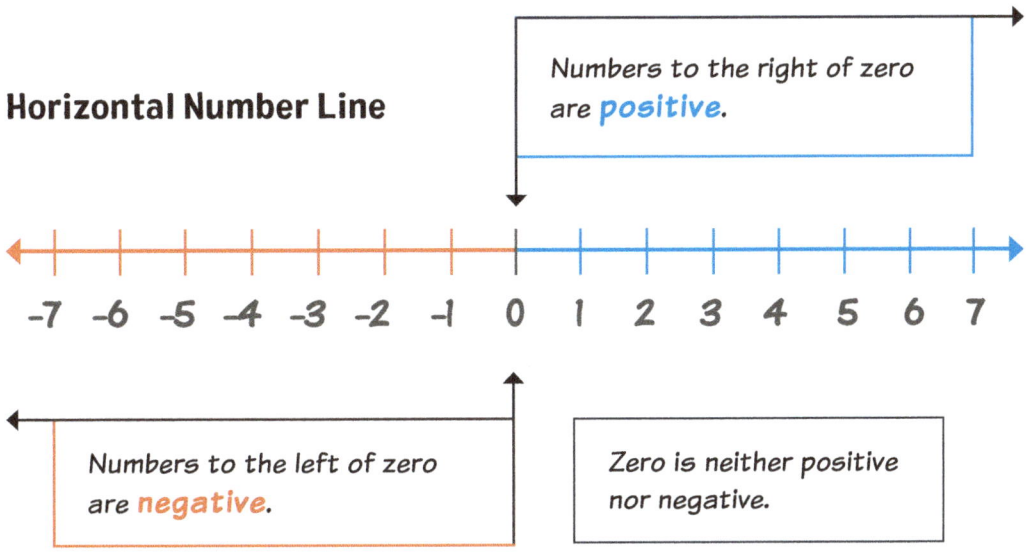

Numbers to the right of zero are *positive*.

Numbers to the left of zero are *negative*.

Zero is neither positive nor negative.

Vertical Number Line

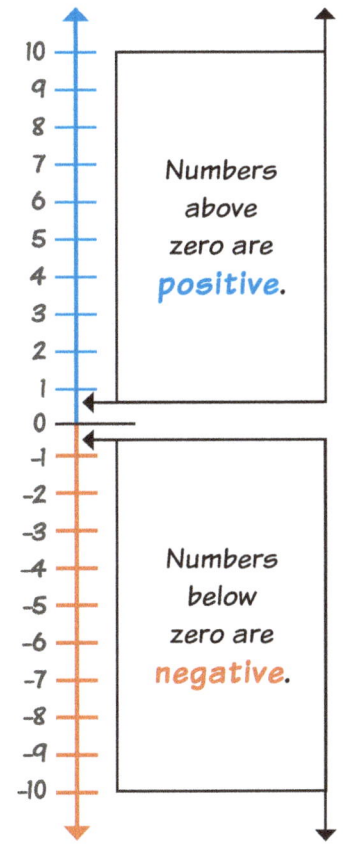

Numbers above zero are *positive*.

Numbers below zero are *negative*.

Comparing Integers

- Which is greater, -4 or -7?

Locate both numbers on a number line. Numbers to the right are greater.

-4 is to the right of -7, so -4 is greater.

- Which is greater, -3 or 4?

Numbers located above are greater.

4 is located above -3, so 4 is greater.

Negative and Positive Numbers (Integers)

Absolute Value

Absolute value is the distance between a number and zero on a number line.

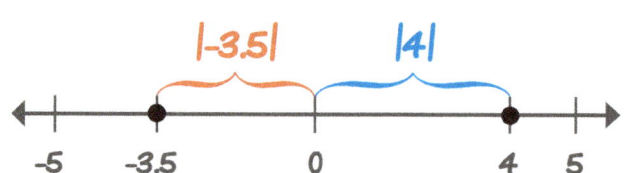

Absolute value is written using parallel vertical lines like this: | |

The **absolute value** of |4| is 4 because it is 4 units away from zero on the number line.

The **absolute value** of |-3.5| is 3.5 because it is 3.5 units away from zero on the number line.

Finding **absolute value** is useful in real life and math. If you want to find how far away you are from the nearest restroom, you don't care if the restroom is behind you or in front of you.

Distance to Restroom	Absolute Value		
-20 yards	|-20|	=	20 yd
60 yards	|60|	=	60 yd

This is the nearest restroom.

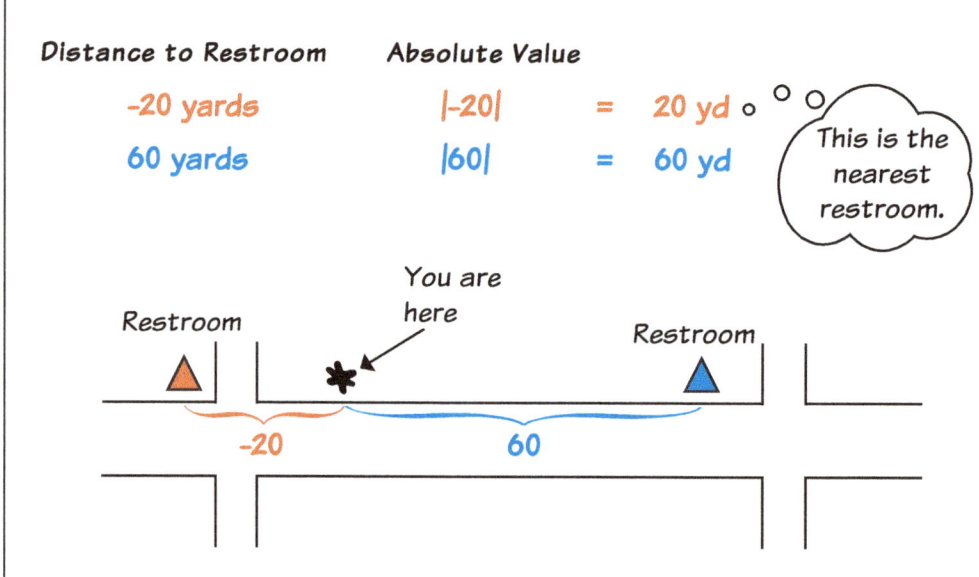

Chapter 8

Fraction Basics

There are many ways to interpret fractions, and in the early years, we tend to focus on the simplest: thinking of a fraction as a part-to-whole ratio. We typically call this "parts of a whole." For many young students, this can be an abstract concept. Understanding that a fraction like $\frac{1}{4}$ means one part of a whole partitioned into four equal parts requires a shift from thinking about whole numbers to thinking about relationships between the number of pieces and the whole.

Even more complex, thinking of $\frac{3}{4}$ of a whole as three $\frac{1}{4}$-sized pieces of a whole partitioned into four pieces presents a new way of considering numbers. The numerator (top number) identifies the number of pieces, and the denominator (bottom number) identifies what kinds of pieces they are.

Fractions are first introduced in the primary grades, when students partition shapes into equal-sized parts, or "fair shares." A square can be partitioned into two equal-sized triangles, a regular pentagon into five. Note that students primarily work with fractions using manipulatives and drawings at this level.

In the early years, fraction work begins early with visual models (shapes, candy bars, etc.) as well as with the written words. For example, we use the word "half" rather than the symbol $\frac{1}{2}$. In most state standards frameworks, the traditional way of writing fractions as symbols doesn't begin until third grade. This is when students are developmentally ready to recognize that the numerator and denominator of a fraction represent a single number on the number line, not two numbers separated by a bar called a "vinculum."

Just as with whole numbers, fractions can be

1. decomposed into smaller parts (e.g., $\frac{3}{4}$ can be decomposed into $\frac{1}{4}+\frac{1}{4}+\frac{1}{4}$),
2. compared to determine the relative size (e.g., $\frac{2}{3}$ is more than $\frac{1}{2}$ and $\frac{2}{5}$ is less than $\frac{1}{2}$, so $\frac{2}{3}$ is greater than $\frac{2}{5}$), and
3. made to represent a division operation (e.g., $12 \div 4$ can be written at $\frac{12}{4}$ and $\frac{1}{3}$ can be written as $1 \div 3$).

A great deal of time is spent on developing these basic ideas of fractions before the middle grades when students learn to combine them, break them apart, or begin to work with ratios expressed as fractions.

It's important to note that across the grades, educators generally agree that there are five interpretations: (1) fractions as parts of wholes or parts of sets, (2) fractions as the result of dividing two numbers, (3) fractions as the ratio of two quantities, (4) fractions as operators, and (5) fractions as measures. More on that in the coming chapters.

Typical Trajectory in Most State Standards Frameworks:

- Grades 1–2: Partition shapes into halves, thirds, and fourths
- Grade 3: Partition shapes into halves, thirds, fourths, fifths, sixths, and eighths; represent fractions on number lines; represent fractions as parts of a set; explore equivalent fractions
- Grade 4: Work with fractions as parts of a whole, parts of a set—denominators of 2, 3, 4, 5, 6, 8, 10, and 100; represent decimals through hundredths in fraction form; represent fractions on a number line and on a ruler
- Grades 5–6: Continue the work with previous grades, working with fractions with any number in the denominator; find equivalent fractions

Partitioned Shapes

Before formal instruction in fractions begins, fraction ideas are developed using shapes and geometry.

Dividing a Circle into Equal Shares (Parts)

This is a whole circle.

This circle is split into **four equal pieces**, called **fourths** or **quarters**.

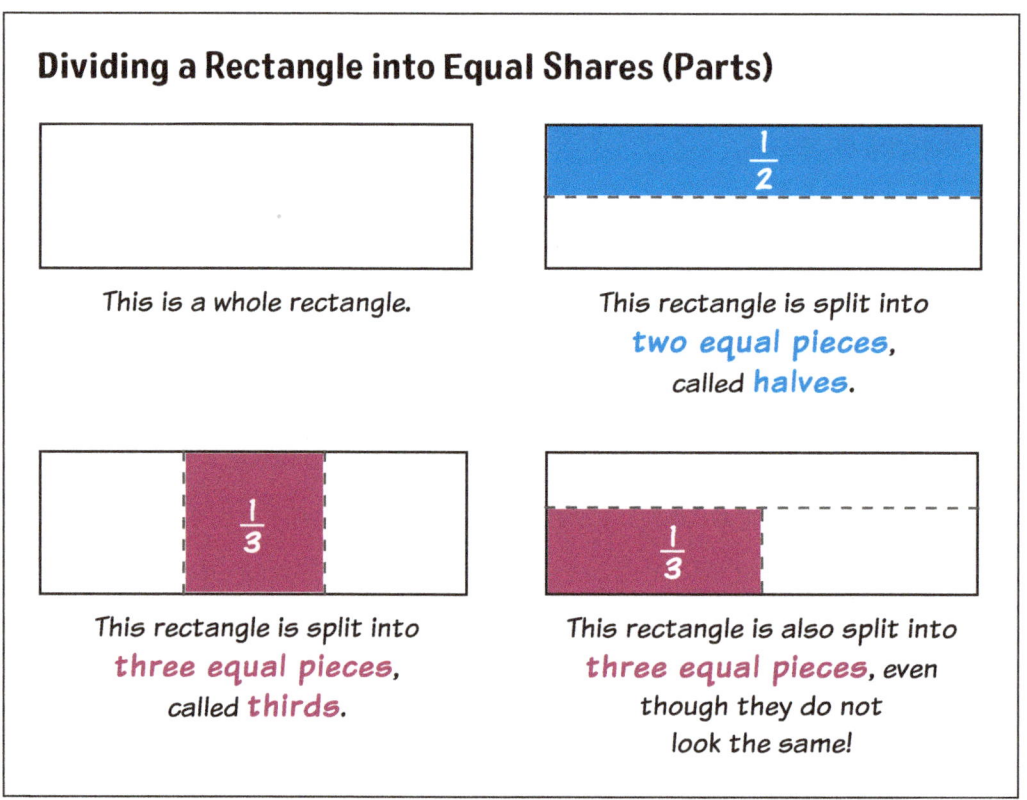

Dividing a Rectangle into Equal Shares (Parts)

This is a whole rectangle.

This rectangle is split into **two equal pieces**, called **halves**.

This rectangle is split into **three equal pieces**, called **thirds**.

This rectangle is also split into **three equal pieces**, even though they do not look the same!

 Students use the words "**fourths**," "**halves**," and "**thirds**," but typically don't use fraction notation ($\frac{1}{4}$, $\frac{1}{2}$, $\frac{1}{3}$, etc.) until third grade.

Fraction Symbols and Representations.........

A **fraction** is a number that represents part of a whole or part of a group. The **denominator** (bottom number) tells how many equal parts are in the whole or set. The **numerator** (top number) tells how many parts you are talking about.

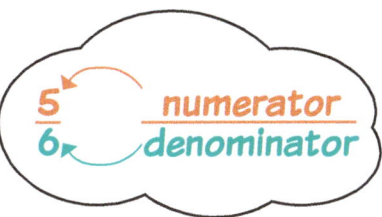

Fractions can be used and named in different ways. Below you will find a few of them:

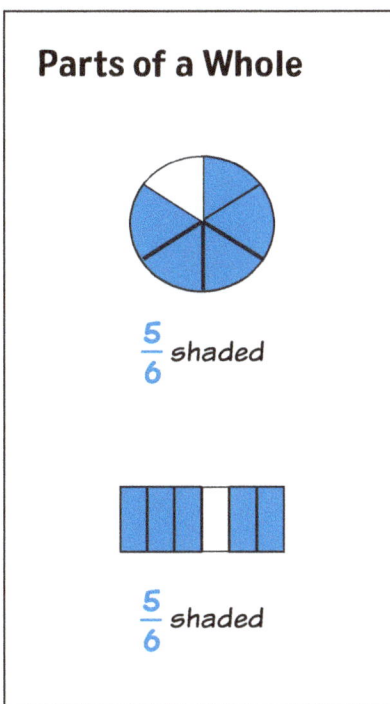

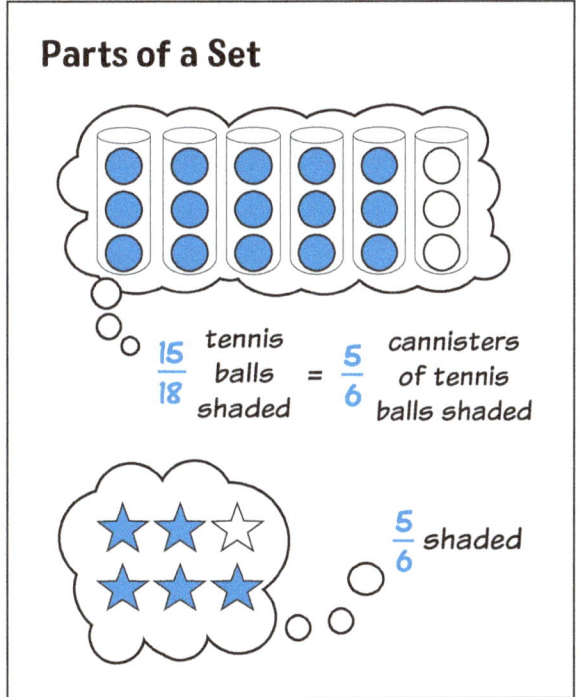

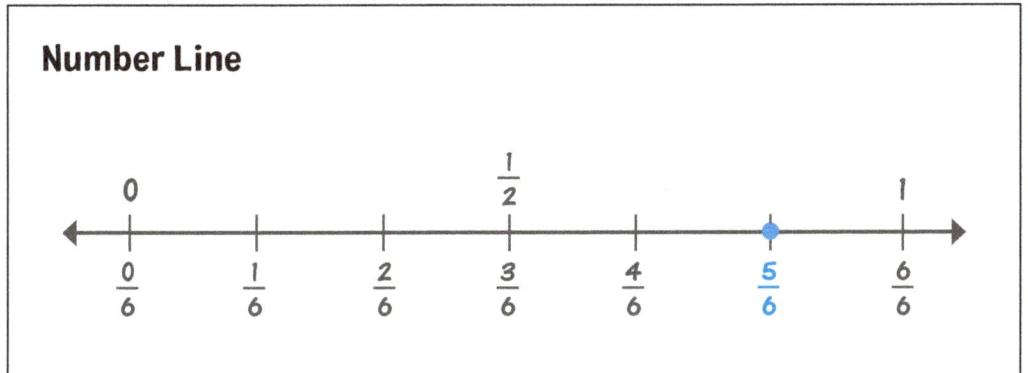

Fraction Basics   **97**

Unit Fractions

A **unit fraction** is a fraction with a numerator of 1. Some examples of unit fractions:

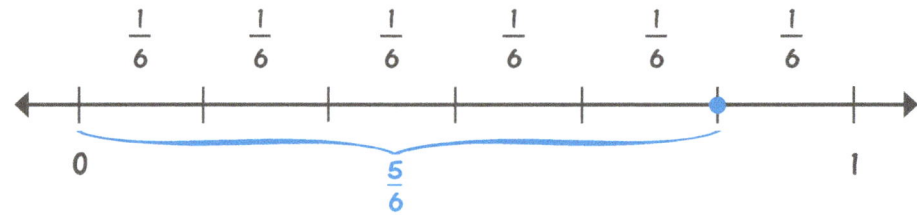

Using Unit Fractions

Fractions with a numerator greater than 1 can be thought of as "multiple copies" of the same fractional piece.

$\frac{5}{6}$ is **5** copies of $\frac{1}{6}$

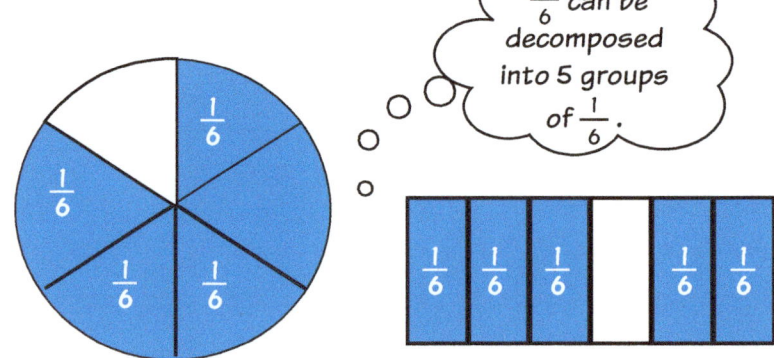

$\frac{5}{6}$ can be decomposed into 5 groups of $\frac{1}{6}$.

For all these examples:

$$\frac{1}{6} + \frac{1}{6} + \frac{1}{6} + \frac{1}{6} + \frac{1}{6} = \frac{5}{6}$$

Decomposing Fractions

Fractions can be decomposed into the sum of their parts, just like whole numbers.

Decomposing a Whole Number	Decomposing a Fraction
13	$\frac{3}{4}$
10 | 3	$\frac{2}{4}$ | $\frac{1}{4}$
13 is the same value as 10 + 3. 13 = 10 + 3	$\frac{3}{4}$ is the same value as $\frac{2}{4} + \frac{1}{4}$. $\frac{3}{4} = \frac{2}{4} + \frac{1}{4}$
5 | 5 | 3	$\frac{1}{4}$ | $\frac{1}{4}$ | $\frac{1}{4}$
13 is the same value as 5 + 5 + 3. 13 = 5 + 5 + 3	$\frac{3}{4}$ is the same value as $\frac{1}{4} + \frac{1}{4} + \frac{1}{4}$. $\frac{3}{4} = \frac{1}{4} + \frac{1}{4} + \frac{1}{4}$

Fractions for Whole Numbers

All whole numbers can be written as **fractions**.

Fractions for One

A fraction is equivalent to one whole when the numerator (top number) and denominator (bottom number) are the same.

$$\frac{5}{5} = 1 \text{ whole} \rightarrow$$

$$\frac{4}{4} = 1 \text{ whole} \rightarrow$$

$$\frac{3}{3} = 1 \text{ whole} \rightarrow$$

Fractions for Other Whole Numbers

Any whole number can be written as a fraction when the numerator is greater than the denominator AND the numerator is a **multiple** of the denominator.

$$3 = \frac{3}{1}$$

A whole is one piece... I have three whole pieces.

$$2 = \frac{6}{3} \qquad 4 = \frac{4}{1} \qquad 3 = \frac{9}{3} \qquad 2 = \frac{2}{1}$$

Fraction Basics

Equivalent Fractions

Equivalent fractions name the same amount. To find equivalent fractions, you multiply (or divide) by any fraction that equals 1.

Equivalent Fractions for One

$$1 = \frac{2}{2} = \frac{3}{3} = \frac{4}{4} = \frac{5}{5} = \frac{17}{17} \ldots$$

Any fraction with the same numerator and denominator equals 1!

Finding Equivalent Fractions by Multiplying

Find equivalent fractions for $\frac{1}{2}$

① ②
$$\frac{1}{2} \times \frac{3}{3} = \frac{1 \times 3}{2 \times 3} = \frac{3}{6}$$
③

① ②
$$\frac{1}{2} \times \frac{4}{4} = \frac{1 \times 4}{2 \times 4} = \frac{4}{8}$$
③

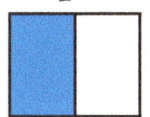 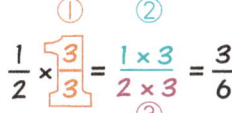 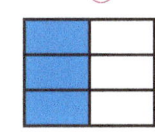

The amount doesn't change, just the size of the pieces!

① Choose any fraction equal to 1.
② Multiply the first numerator by the second numerator.
③ Multiply the first denominator by the second denominator.

Finding Equivalent Fractions by Dividing

Find equivalent fractions for $\frac{6}{12}$

① ②
$$\frac{6}{12} \div \frac{3}{3} = \frac{6 \div 3}{12 \div 3} = \frac{2}{4}$$
③

① ②
$$\frac{6}{12} \div \frac{2}{2} = \frac{6 \div 2}{12 \div 2} = \frac{3}{6}$$
③

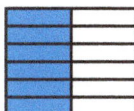 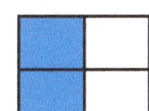 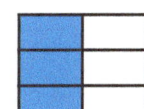

① Choose any fraction equal to 1.
② Divide the first numerator by the second numerator.
③ Divide the first denominator by the second denominator.

Fraction Basics

Comparing Fractions With the Same Denominator

Fractions split into pieces of the same size can be compared directly.

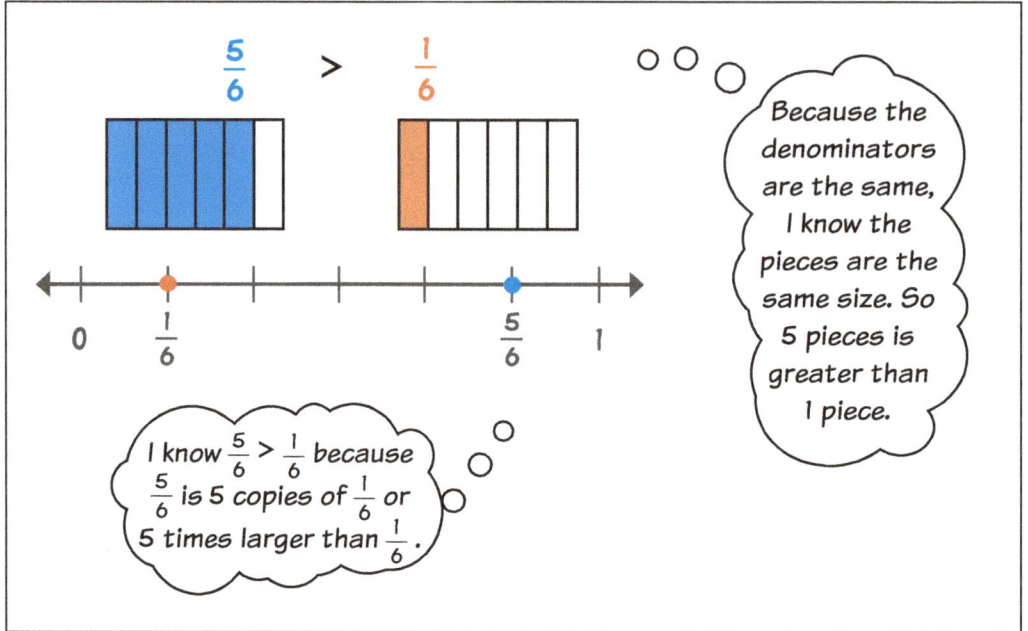

Comparing Fractions With the Same Numerator

Fractions with the same number of pieces (even if the sizes of the pieces differ) can be compared directly.

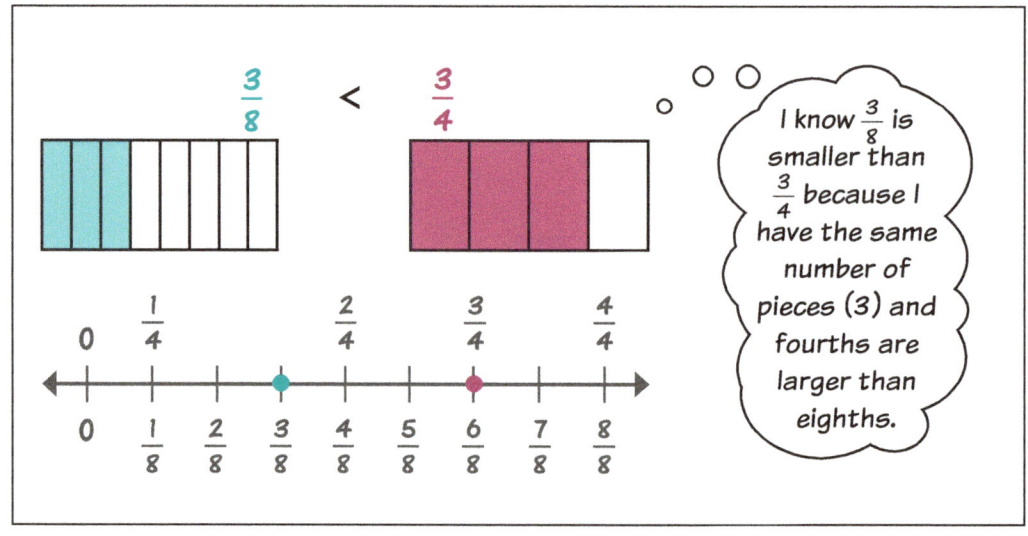

Comparing Fractions With Different Denominators and Numerators

You can compare fractions with different denominators by using **benchmark fractions** or **finding equivalent fractions**. Like a benchmark number, a **benchmark fraction** is an easy fraction to work with, like $0, \frac{1}{4}, \frac{1}{2}, \frac{3}{4}$, or 1.

Using Benchmark Fractions

Which is larger $\frac{3}{4}$ or $\frac{5}{6}$?

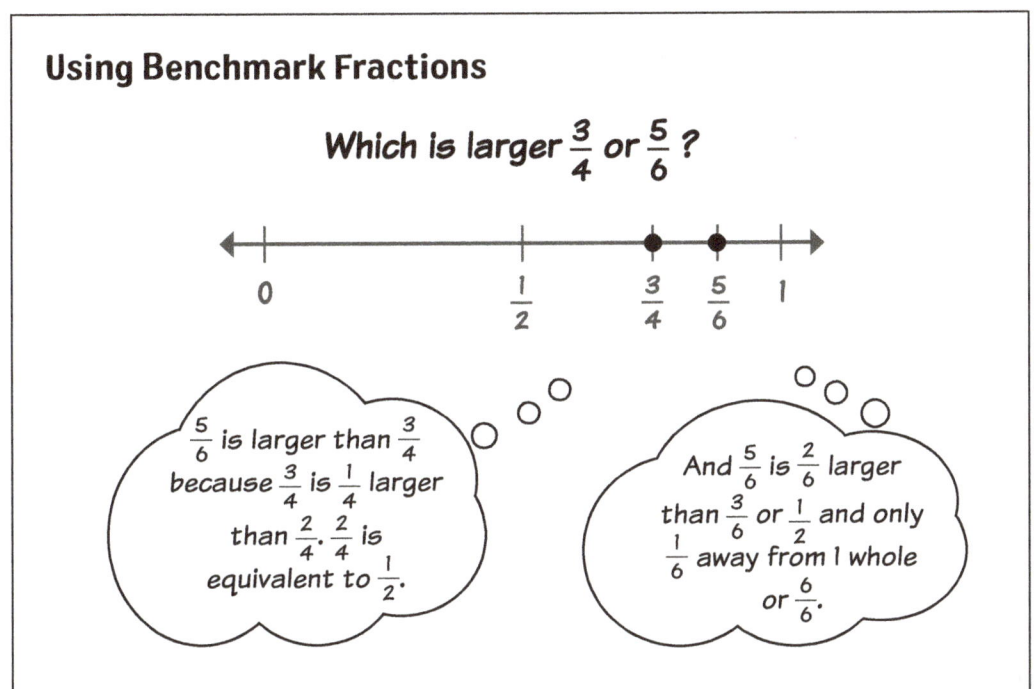

$\frac{5}{6}$ is larger than $\frac{3}{4}$ because $\frac{3}{4}$ is $\frac{1}{4}$ larger than $\frac{2}{4}$. $\frac{2}{4}$ is equivalent to $\frac{1}{2}$.

And $\frac{5}{6}$ is $\frac{2}{6}$ larger than $\frac{3}{6}$ or $\frac{1}{2}$ and only $\frac{1}{6}$ away from 1 whole or $\frac{6}{6}$.

Finding Equivalent Fractions

Which is larger $\frac{3}{4}$ or $\frac{5}{6}$?

$\frac{3}{4} \times \frac{3}{3} = \frac{3 \times 3}{4 \times 3} = \frac{9}{12}$ $\frac{5}{6} \times \frac{2}{2} = \frac{5 \times 2}{6 \times 2} = \frac{10}{12}$

$\frac{9}{12} < \frac{10}{12}$

$\frac{3}{4} < \frac{5}{6}$

I'll multiply each fraction by a whole number fraction equal to 1 that will make my denominators the same, so it will be easier to compare them.

Fraction Basics

Finding Common Denominators

To compare, add or subtract fractions with unlike denominators, you may want to rename the fractions so that they have **common denominators**.

Step 1

Find the *least common multiple** for the **denominators**.

multiples of 3

$\frac{2}{3} \rightarrow$ 0, 3, 6, 9, 12, 15

$\frac{1}{2} \rightarrow$ 0, 2, 4, 6, 8, 10

multiples of 2

*See page 82 for more about least common multiples.

Step 2

Create equivalent fractions by multiplying by a fraction equal to 1 that will make both **denominators** equal to the *least common multiple*.

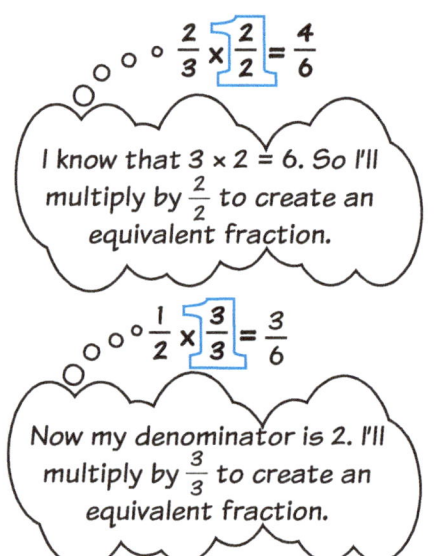

$\frac{2}{3} \times \frac{2}{2} = \frac{4}{6}$

I know that 3 × 2 = 6. So I'll multiply by $\frac{2}{2}$ to create an equivalent fraction.

$\frac{1}{2} \times \frac{3}{3} = \frac{3}{6}$

Now my denominator is 2. I'll multiply by $\frac{3}{3}$ to create an equivalent fraction.

Step 3

Use the fractions with **like denominators** to compare, add, or subtract.

Compare: $\quad \frac{4}{6} > \frac{3}{6} \quad$ so $\quad \frac{2}{3} > \frac{1}{2}$

Add: $\quad \frac{4}{6} + \frac{3}{6} = \frac{7}{6} = 1\frac{1}{6} \quad$ so $\quad \frac{2}{3} + \frac{1}{2} = 1\frac{1}{6}$

Subtract: $\quad \frac{4}{6} - \frac{3}{6} = \frac{1}{6} \quad$ so $\quad \frac{2}{3} - \frac{1}{2} = \frac{1}{6}$

Chapter 9

Add, Subtract, Multiply, and Divide Fractions

The use of physical and visual models is crucial when students are initially adding, subtracting, multiplying, and dividing fractions. Concrete and pictorial representations allow learners to "see" the value of these numbers and promote deeper understanding of how fractions relate to each other as well as how to combine them.

Students often "discover" the general rules for working with fractions from the strategic use of appropriate models. As teachers, we should be cautious not to push algorithms before this visual understanding is nurtured.

Be sure to use a variety of representations to model fractions. Too often, students only shade in shapes such as circles and squares to represent fractional parts. This will hinder the trajectory for later years. Be sure students also model fractions on number lines and as parts of sets. This helps students visualize comparisons in really helpful ways. For example, if a number line is drawn with a point to represent $\frac{1}{4}$, students can easily recognize that it is closer to zero than it is to one whole.

Multiplying fractions also becomes clearer with a number line. Imagine multiplying the same $\frac{1}{4}$ five times. When a segment representing $\frac{1}{4}$ is drawn and then iterated five times, students can see that the new segment is $\frac{5}{4}$ units long, which is the same as $1\frac{1}{4}$ units long.

Intentional work with these and other visuals such as area models and fraction strips encourage a deeper understanding of how fractions relate to each other and how to manipulate them before moving into more traditional algorithms.

Typical Trajectory in Most State Standards Frameworks:

- Grade 3: Beginning fraction addition and subtraction using conceptual strategies
- Grade 4: Fraction addition and subtraction using conceptual strategies; fraction multiplication using conceptual strategies
- Grade 5: Fraction addition and subtraction; fraction multiplication; fraction division using conceptual strategies
- Grade 6: All fraction operations

Adding and Subtracting Fractions

Video 11: Fraction Pieces: You can use commercial or home-made fraction pieces to demonstrate fraction concept and operations. *Pictured here: fraction circles and foam fraction bars, fraction squares, fraction towers, and pattern blocks, Cuisenaire® rods*

https://qrs.ly/svg99le

To add and subtract **fractions**, the **denominators** (bottom numbers) must be the same. Then add or subtract the **numerators** (top numbers).

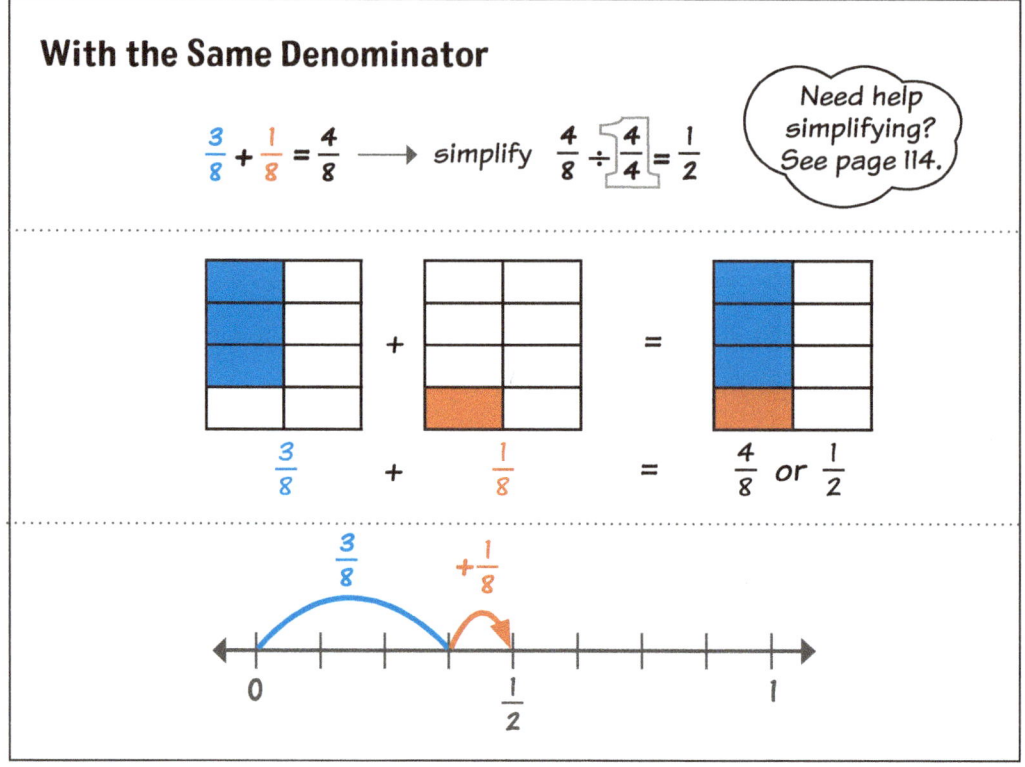

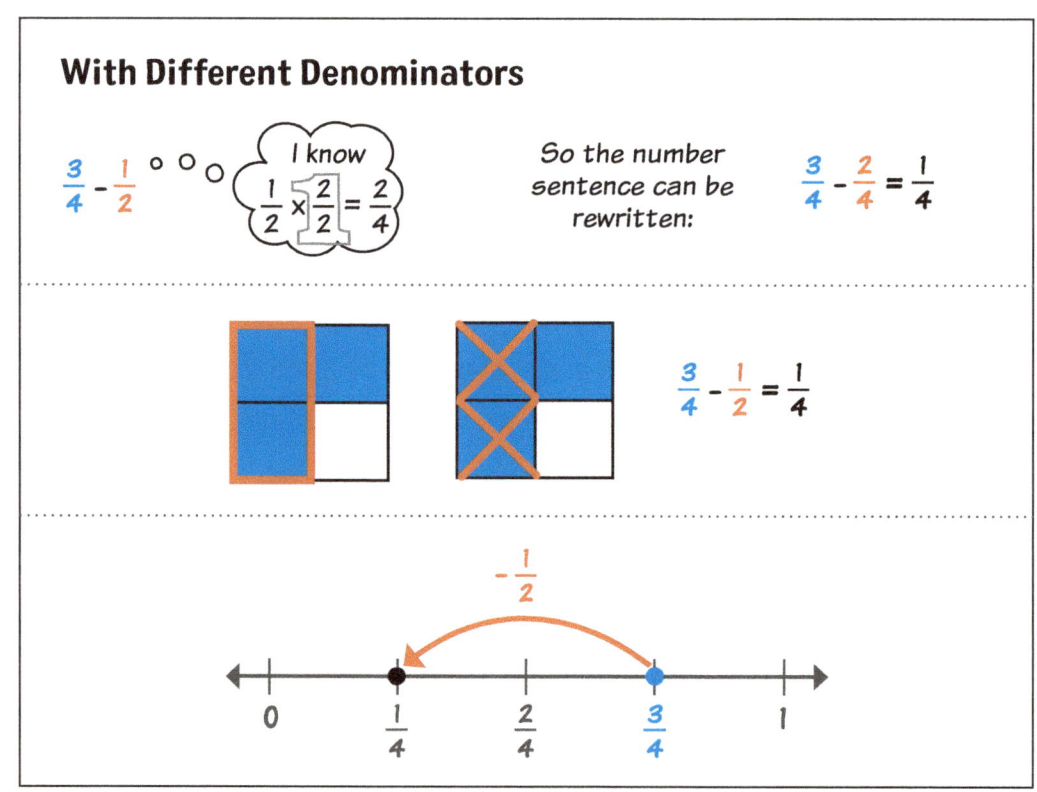

Adding and Subtracting Mixed Numbers........

When adding or subtracting **mixed numbers**, your strategy may depend on whether the two numbers share a common denominator.

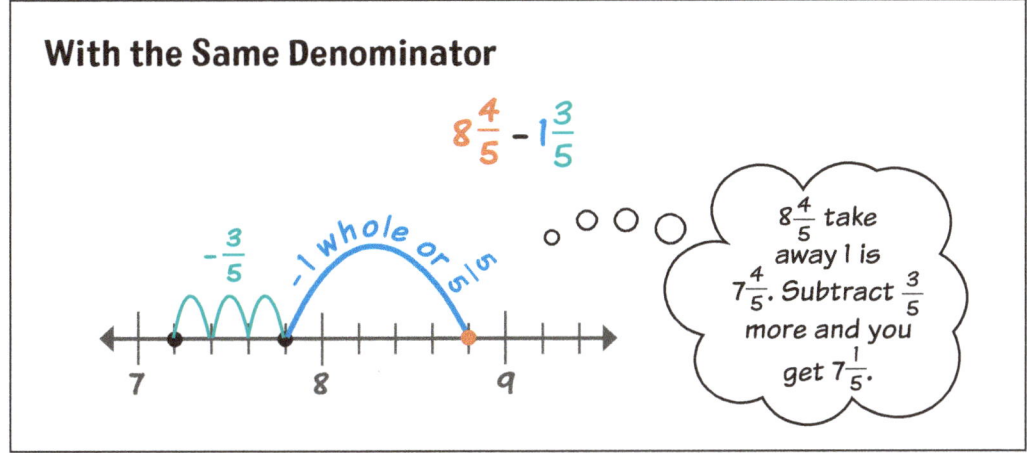

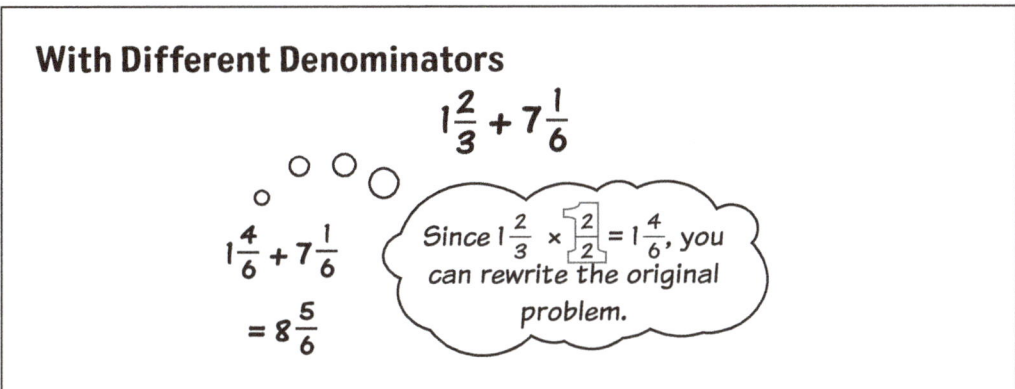

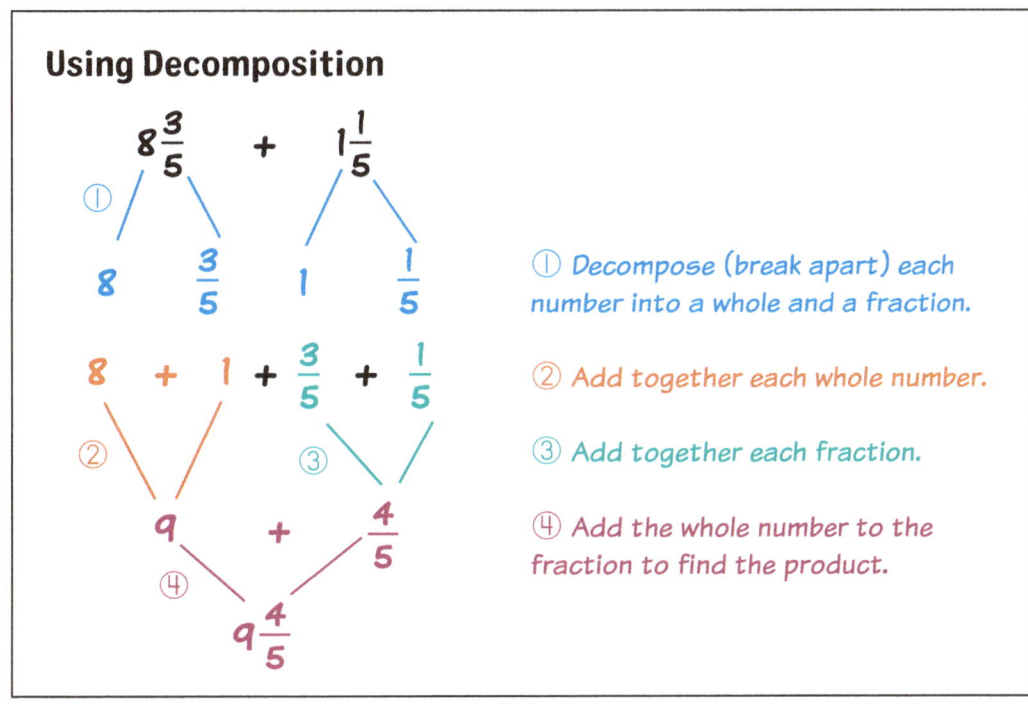

Add, Subtract, Multiply, and Divide Fractions

Solving Addition and Subtraction Word Problems Involving Fractions

I bought $2\frac{1}{2}$ gallons of paint. After painting my room, I still have $\frac{1}{4}$ of a gallon remaining. How much paint did I use?

$2\frac{1}{2} = 2\frac{2}{4}$

I know $\frac{1}{2}$ is the same as $\frac{2}{4}$.

$2\frac{2}{4} - \frac{1}{4} = 2\frac{1}{4}$ gallons

I used $2\frac{1}{4}$ gallons of paint.

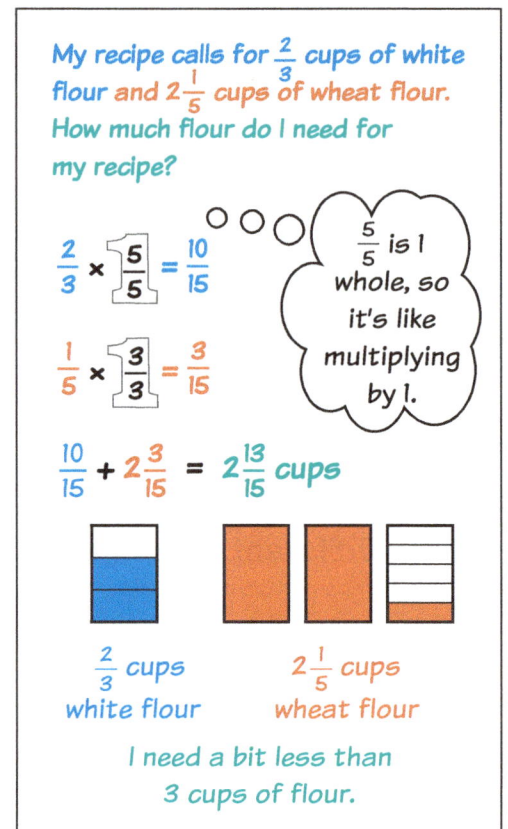

My recipe calls for $\frac{2}{3}$ cups of white flour and $2\frac{1}{5}$ cups of wheat flour. How much flour do I need for my recipe?

$\frac{2}{3} \times \frac{5}{5} = \frac{10}{15}$

$\frac{1}{5} \times \frac{3}{3} = \frac{3}{15}$

$\frac{5}{5}$ is 1 whole, so it's like multiplying by 1.

$\frac{10}{15} + 2\frac{3}{15} = 2\frac{13}{15}$ cups

$\frac{2}{3}$ cups white flour $2\frac{1}{5}$ cups wheat flour

I need a bit less than 3 cups of flour.

 Sometimes when solving word problems with fractions it is necessary to carefully consider the unit quantities. **This is a common student error.**

Bill had $\frac{2}{3}$ of a cup of juice. He drank $\frac{1}{2}$ of his juice.

How much juice did Bill have left?

Bill has $\frac{2}{3}$ of a cup. { He drank $\frac{1}{2}$ of his juice.
Bill has $\frac{1}{2}$ of $\frac{2}{3}$ or $\frac{1}{3}$ of the whole cup left.

This problem cannot be solved by subtracting $\frac{2}{3} - \frac{1}{2}$ because the $\frac{2}{3}$ refers to a cup of juice, and $\frac{1}{2}$ refers to the amount of juice Bill had and not to the whole cup of juice.

Solving Multiplication and Division Word Problems Involving Fractions.................

	Multiplication	**Division**
Equal Groups of Objects	Grandma's recipe for banana muffins calls for $\frac{3}{4}$ cup of oats. You are making $\frac{1}{2}$ of the recipe. How many cups of oats should you use? $\frac{1}{2} \cdot \frac{3}{4} = \boxed{}$ $\left(\frac{1}{2} \cdot \frac{3}{4} = \frac{3}{8} \text{ cup}\right)$	Molly is following a lemonade recipe that calls for $2\frac{1}{2}$ cups of sugar. If she finds 10 cups of sugar in the pantry, how many batches of lemonade can she make? $10 \div 2\frac{1}{2} = \boxed{}$ $2\frac{1}{2} \times \boxed{} = 10$ $\left(10 \div 2\frac{1}{2} = 4 \text{ batches}\right)$
Arrays of Objects	Fiona is arranging her stamp collection into rows. If 1 stamp is $\frac{3}{4}$ of an inch wide, how long will a row of 12 stamps be? $\frac{3}{4} \cdot 12 = \boxed{}$ $\left(\frac{3}{4} \cdot 12 = 9 \text{ inches}\right)$	A small T-shirt requires $\frac{3}{4}$ yards of fabric. How many T-shirts can be made from 48 yards? $48 \div \frac{3}{4} = \boxed{}$ $\frac{3}{4} \cdot \boxed{} = 48$ $\left(48 \div \frac{3}{4} = 64 \text{ T-shirts}\right)$
Compare	A father weighs $2\frac{1}{4}$ times more than his son. If the son weighs 60 pounds, how much does the father weigh? $60 \cdot 2\frac{1}{4} = \boxed{}$ $\left(60 \cdot 2\frac{1}{4} = 135 \text{ pounds}\right)$	A new Razor scooter is on sale for $60.00. Its regular price is $90.00. What fraction of the regular price is the sale price? $60 \div 90 = \boxed{}$ $\left(60 \div 90 = \frac{2}{3}\right)$

Add, Subtract, Multiply, and Divide Fractions

Multiplying Fractions by Whole Numbers........

When you **multiply** a whole number by a fraction that is less than 1, the **product** is less than the original whole number because you're only taking a part of it.

- "x" means "of" or "groups of."
- 3 × 4 means "3 groups of 4."
- $\frac{2}{3}$ × 15 means "$\frac{2}{3}$ of 15."

What is $\frac{2}{3}$ of 15? $\frac{2}{3}$ × 15 = ☐

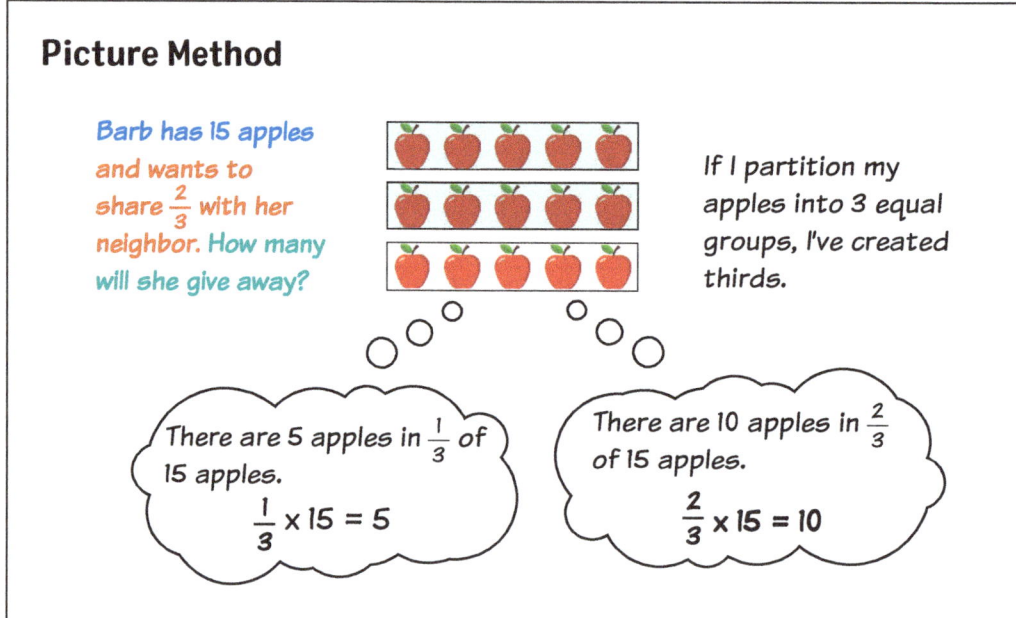

Traditional Method

$$\frac{2}{3} \times 15 = \frac{2}{3} \times \frac{15}{1} = \frac{2 \times 15}{3 \times 1} \rightarrow \frac{30}{3} = 10$$

with steps labeled ① ② ③ ④

① Write the whole number as a fraction over 1.

② Multiply the numerators (top); the result is your new numerator.

③ Multiply the denominators (bottom); the result is your new denominator.

④ Rename as a simpler equivalent fraction or whole number, if possible.

Fraction Operations

112   SEEING THE MATH YOU TEACH, GRADES K–6

Multiplying Fractions by Fractions

When multiplying fractions, it is sometimes helpful to start with a visual representation.

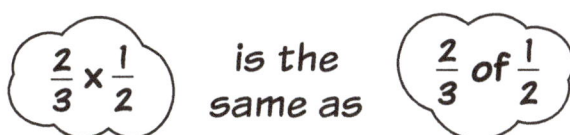

Picture Method

Start with $\frac{1}{2}$ and partition that piece into **thirds**.

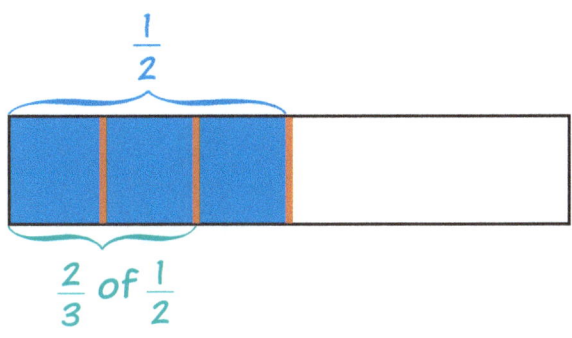

$\frac{2}{3}$ **of the half** you started with is equal to **2 of 3 pieces in the half**, which would fit into the whole $\frac{2}{3} \times \frac{1}{2} = \frac{2}{6}$.

Rename your fraction to a simpler equivalent fraction:

$$\frac{2}{6} \div \frac{2}{2} = \frac{1}{3}$$

Traditional Method

$$\frac{2}{3} \times \frac{1}{2} = \frac{2 \times 1}{3 \times 2} = \frac{2}{6} = \frac{1}{3}$$

① Multiply the numerators (top); the result is your new numerator.
② Multiply the denominators (bottom); the result is your new denominator.
③ Rename as a simpler equivalent fraction, if possible.

Add, Subtract, Multiply, and Divide Fractions

Multiplying Fractions With Mixed Numbers......

There are a few different strategies for **multiplying fractions with mixed numbers**. Below are three of them:

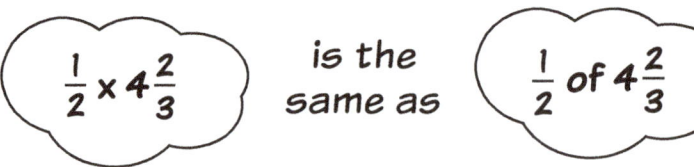

$\frac{1}{2} \times 4\frac{2}{3}$ is the same as $\frac{1}{2}$ of $4\frac{2}{3}$

Picture Method

Start with $4\frac{2}{3}$ and determine what half would be:

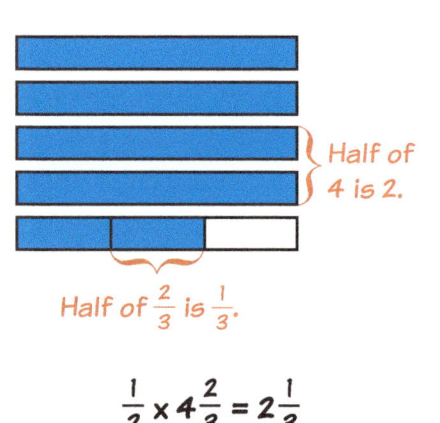

Half of 4 is 2.

Half of $\frac{2}{3}$ is $\frac{1}{3}$.

$$\frac{1}{2} \times 4\frac{2}{3} = 2\frac{1}{3}$$

Traditional Method

First, convert the mixed number into an improper fraction:

$$\frac{1}{2} \times 4\frac{2}{3}$$

$$\frac{1}{2} \times \left(\frac{12}{3} + \frac{2}{3}\right)$$

$$\frac{1}{2} \times \frac{14}{3}$$

$$\frac{1 \times 14}{2 \times 3} = \frac{14}{6}$$

Since I am working with thirds, one whole is $\frac{3}{3}$. I have 4 wholes $\frac{3}{3} + \frac{3}{3} + \frac{3}{3} + \frac{3}{3}$ or $\frac{12}{3}$.

Then, rename your answer from an improper fraction back to a **mixed number**:

Simplify!

$$\frac{14}{6} = \frac{6}{6} + \frac{6}{6} + \frac{2}{6} = 2\frac{2}{6} = 2\frac{1}{3}$$

Decomposition Method*

$$\frac{1}{2} \times 4\frac{2}{3}$$

$$\left(\frac{1}{2} \times 4\right) + \left(\frac{1}{2} \times \frac{2}{3}\right)$$

$$2 + \frac{2}{6}$$

$$= 2\frac{2}{6}$$

*See page 115 for steps.

Fractions aren't always renamed to simpler equivalent fractions.

Multiplying Fractions by Fractions Using Decomposition

You can multiply two fractions using decomposition, just as you would multiply two whole numbers (see page 47).

Colin and his dad finished a building project. They had $2\frac{1}{2}$ boards left, and each board was originally $4\frac{2}{3}$ feet long. In all, how many feet of board did they have left?

$2\frac{1}{2} \times 4\frac{2}{3}$

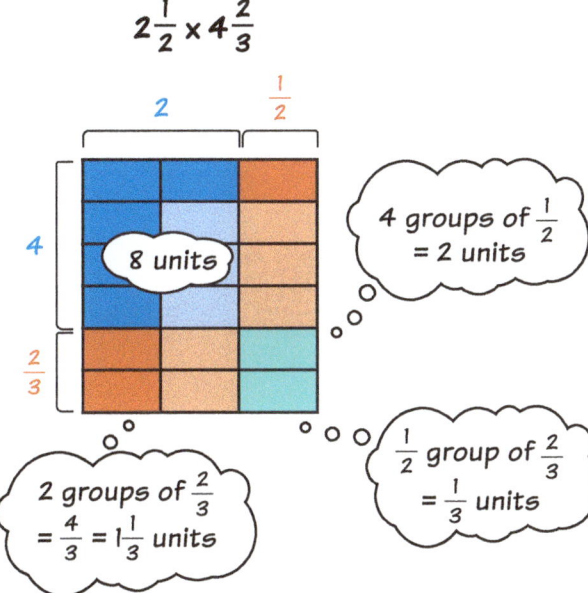

4 groups of $\frac{1}{2}$ = 2 units

2 groups of $\frac{2}{3}$ = $\frac{4}{3}$ = $1\frac{1}{3}$ units

$\frac{1}{2}$ group of $\frac{2}{3}$ = $\frac{1}{3}$ units

You can use an area model to multiply fractions.

① Represent $2\frac{1}{2}$ as 2 wholes and $\frac{1}{2}$ horizontally. Represent $4\frac{2}{3}$ as 4 wholes and $\frac{2}{3}$ vertically on the same area model.

② Fill in the area to complete the rectangle.

③ Add up the number of units from each creation to calculate the product.

$8 + 2 + 1\frac{1}{3} + \frac{1}{3}$

$11\frac{2}{3}$ feet

$2\frac{1}{2} \times 4\frac{2}{3}$

2   $\frac{1}{2}$ 4 $\frac{2}{3}$

$(2 \times 4) + (2 \times \frac{2}{3}) + (\frac{1}{2} \times 4) + (\frac{1}{2} \times \frac{2}{3})$

$8 \quad + \quad \frac{4}{3} \quad + \quad 2 \quad + \quad \frac{1}{3}$

$8 + 2 = 10 \qquad \frac{4}{3} + \frac{1}{3} = \frac{5}{3}$ or $1\frac{2}{3}$

$10 + 1\frac{2}{3} = 11\frac{2}{3}$

① Decompose (break apart) each number into a whole and a fraction.

② Multiply each part of the first number by each part of the second number.

③ Add together each partial product.

④ Now you have the answer (product)!

Add, Subtract, Multiply, and Divide Fractions

Division With Fractions

Video 12: Division With Fraction Pieces: You can use physical or virtual fraction pieces to demonstrate fraction division. *Pictured here: fraction square created in a slide deck*

https://qrs.ly/ogg99lh

When you **divide fractions** you are determining how many times the second number (divisor) "fits into" the first number (dividend), just like when you divide whole numbers.

| dividend ÷ divisor = quotient |

Whole Number ÷ Whole Number

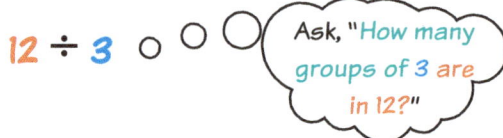

Ask, "How many groups of 3 are in 12?"

Picture Method

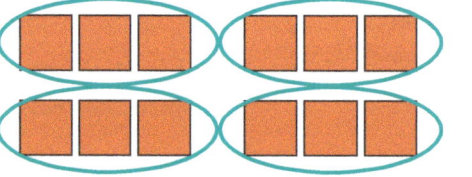

There are **four** groups of 3 in 12.

Whole Number ÷ Fraction

A doll's dress requires $\frac{1}{2}$ a yard of fabric. How many dresses can I make using 4 yards?

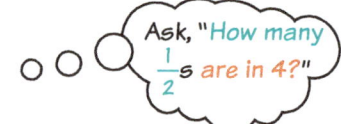

Ask, "How many $\frac{1}{2}$s are in 4?"

Picture Method

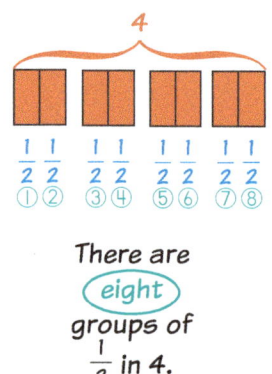

There are **eight** groups of $\frac{1}{2}$ in 4.

Common Denominator Method

$$4 \div \frac{1}{2} = \frac{8}{2} \div \frac{1}{2} = \frac{8 \div 1}{2 \div 2} = \frac{8}{1} = 8$$

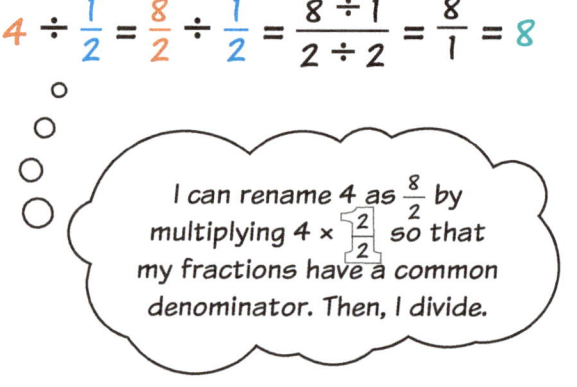

I can rename 4 as $\frac{8}{2}$ by multiplying $4 \times \frac{2}{2}$ so that my fractions have a common denominator. Then, I divide.

Division With Fractions (continued)

Fraction ÷ Fraction

A cookie recipe needs $\frac{1}{8}$ pounds of butter. I have $\frac{3}{4}$ pounds of butter. How many batches of cookies can I make with the butter I have?

$$\frac{3}{4} \div \frac{1}{8}$$

Ask "How many $\frac{1}{8}$s are in $\frac{3}{4}$?"

Picture Method

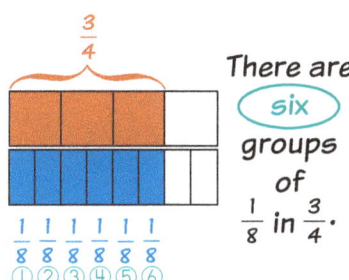

There are **six** groups of $\frac{1}{8}$ in $\frac{3}{4}$.

Common Denominator Method*

$$\frac{3}{4} \div \frac{1}{8} = \frac{6}{8} \div \frac{1}{8} = \frac{6 \div 1}{8 \div 8} = \frac{6}{1} = 6$$

*See page 104 for more help finding common denominators.

Fraction ÷ Larger Fraction

A cookie recipe needs $\frac{1}{2}$ of a pound of chocolate chips. I have $\frac{1}{4}$ of a pound of chocolate chips. How many batches of cookies can I make with the chips I have?

$$\frac{1}{4} \div \frac{1}{2}$$

Ask, "How many $\frac{1}{2}$s are in $\frac{1}{4}$?"

Picture Method

This represents having $\frac{1}{4}$ of the square.

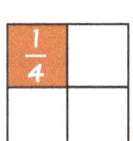

This represents dividing the square into pieces that are each $\frac{1}{2}$ of the square.

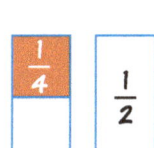

How many pieces of size $\frac{1}{2}$ of the square are in each $\frac{1}{4}$ of the square?

There is $\frac{1}{2}$ of a piece of $\frac{1}{2}$ of the square in $\frac{1}{4}$ of the square.

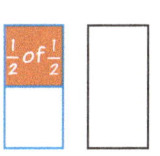

Reciprocal Multiplication Method

$$\frac{1}{4} \div \frac{1}{2} = \frac{1}{4} \times \frac{2}{1} = \frac{2}{4} = \frac{1}{2}$$

I can divide fractions by multiplying by the reciprocal. For example, the reciprocal of $\frac{1}{2}$ is $\frac{2}{1}$ or 2. The numerator (top) and denominator (bottom) flipped places.

Add, Subtract, Multiply, and Divide Fractions

Chapter 10

Relationships Between Fractions, Decimals, and Percents

Fractions, decimals, and percents can be used to represent the same number in different forms. The ability to switch between these forms allows for a deeper understanding of numerical relationships, enables easier comparisons between these values, and is crucial for performing calculations in real-world contexts.

Fourth- and fifth-graders are expected to see the relationship between fractions and decimals both in their symbolic representations and in their values. For example, they should understand that 0.75 can be represented as $\frac{75}{100}$ and as $\frac{3}{4}$. They should also see this connection with all four operations by representing operations in different forms, such as thinking of 0.8×0.3 as $\frac{8}{10} \times \frac{3}{10}$.

Middle school students are expected to recognize different forms of the same number and have the flexibility to switch between the forms that are best for the situation. Of course, fractions are a student's first introduction to abstraction since they are introduced much earlier in the learning trajectory. Once learners develop a solid understanding of fractions, they can extend that understanding to decimals and percents.

Decimal understanding is strongest when it builds on a student's understanding of fractions. For example, when trying to understand why 0.01 (one hundredth) is smaller than 0.1 (one tenth), students must be able to build on their understanding of what one tenth truly is. Working with fractions first helps make this transition smoother.

Thinking of decimal fractions as a subset of all fractions is very helpful for many students. After all, decimal fractions are simply the fractions with powers of ten in the denominators (tenths, hundredths, thousandths, etc.). Furthermore, students benefit from understanding that the decimal form and fraction form of numbers represent the same quantities.

Typical Trajectory in Most State Standards Frameworks:

- Grades K–5: Fractions
- Grade 4: Decimals to hundredths represented in both decimal and fraction form
- Grade 5: Decimals to thousandths represented in both decimal and fraction form
- Grade 6: Percents, decimals, and fractions used to represent common quantities and amounts

Relationships Between Fractions, Decimals, Percents, and Ratios

Any number that can be written as a ratio (comparison of two numbers) with integers (positive or negative whole numbers or zero) is considered a **rational number**. Fractions, decimals and percents all fall into that category.

Ratios

Ratios compare two numbers and are often written in fraction form.

Katie has 30 CDs. Craig has 20 CDs. How much bigger is Katie's collection?

$\frac{\text{Katie's CDs}}{\text{Craig's CDs}} = \frac{30}{20} = 1\frac{1}{2}$ Katie's collection is $1\frac{1}{2}$ times bigger than Craig's.

Ratio in fraction form.

Katie's CDs : Craig's CDs
30 : 20
3 : 2

For every 3 CDs Katie has, Craig has 2.

Ratio using a colon.

Three to two.

Fractions

Fractions are numbers that represent part of a whole or part of a group.

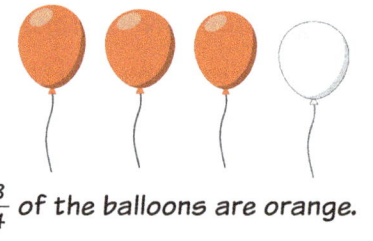

$\frac{3}{4}$ of the balloons are orange.

Source: Balloons image from iStock.com/Photoplotnikov

Percents

The word **percent** comes from a Latin word meaning "by the hundred." In math, a percentage is a fraction with the denominator 100.

$45\% = \frac{45}{100} =$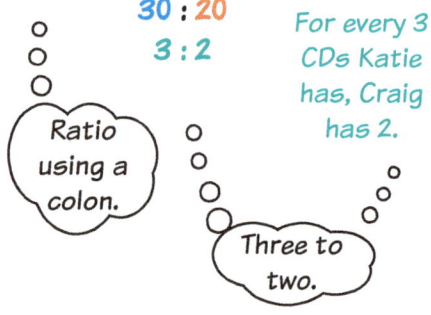

$98\% = \frac{98}{100} =$

Decimals

Fractions that have a special set of denominators (tenths, hundredths, thousandths, . . .) can be written as **decimals**. Instead of being written using a fraction bar, they are written using a decimal point.

$\frac{7}{10}$ = seven tenths = 0.7 $\frac{84}{100}$ = eighty-four hundredths = 0.84

120   SEEING THE MATH YOU TEACH, GRADES K–6

Decimal Fractions

Fractions with a denominator (bottom number) of 10 or 100 are sometimes called **decimal fractions** and are easily converted to decimals or money.

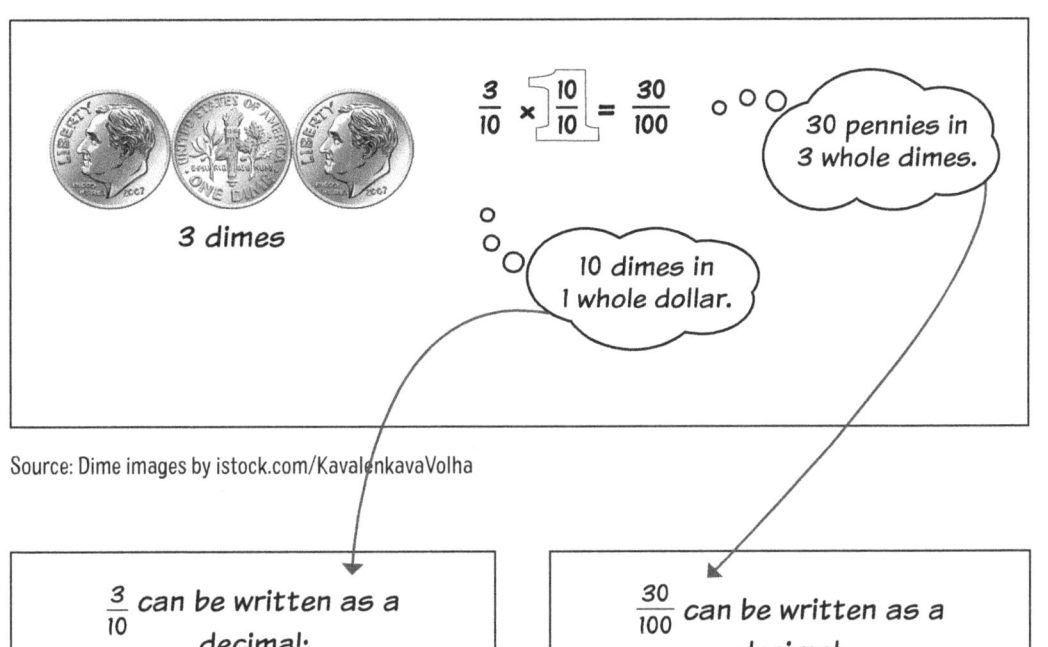

Source: Dime images by istock.com/KavalenkavaVolha

Video 13: Decimal Fractions With Proportional Coin Cards (free download): You can use proportional coin cards to help students connect fraction, decimal, and percent representations. *Pictured here: KP® Dollar Board and coin cards—click on link below to download.*

https://qrs.ly/cug99lp

Link to *KP® Dollar Board and coin cards* download: https://qrs.ly/x5g99ju

---

$\frac{3}{10}$ can be written as a decimal:

0.3

"three tenths"

$\frac{30}{100}$ can be written as a decimal:

0.30

30 cents or $0.30
"thirty hundredths"

$\frac{4}{100}$ can be written as a decimal:

0.04

4 cents or $0.04
"four hundredths"

$\frac{7}{1000}$ can be written as a decimal:

0.007

"seven thousandths"

 It takes 10 pennies to make a dime and 10 dimes to make a dollar.

This 10-to-1 relationship works for all digits that are side-by-side in our place value system as you move from right to left.

Relationships Between Fractions, Decimals, and Percents

Converting Fractions to Decimals and Percents

Fractions can be written as a **decimal** or a **percent**.

One way to convert a fraction to a decimal is by making equivalent fractions with a denominator of 10, 100, 1,000, etc.

$$\frac{1}{2} \times \frac{5}{5} = \frac{5}{10} = 0.5$$

$$\frac{3}{4} \times \frac{25}{25} = \frac{75}{100} = 0.75$$

I know 2 × 5 is 10, so I'll multiply by $\frac{5}{5}$ to change my denominator to 10.

I know 4 × 25 is 100, so I'll multiply by $\frac{25}{25}$ to change my denominator to 100.

Another way to convert a fraction to a decimal is to divide. This works because a fraction is one way to show division.

Think $\frac{1}{8} = 1 \div 8.$

```
   0.125
8)1.000
  -8↓
   20
  -16↓
    40
   -40
     0
```

Percent means "by the 100." You can use fractions or decimals written in hundredths or percents to name the same number.

$$\frac{71}{100} = 0.71 = 71\%$$

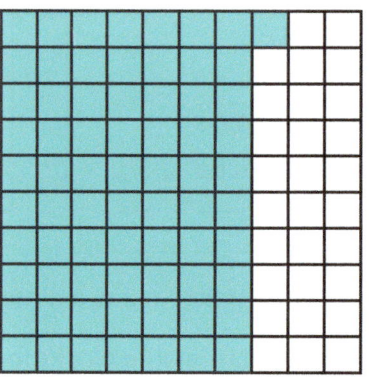

Chapter 11

Ratios and Rates

Ratios and rates are essential in middle grades mathematics because they form a foundational concept for understanding proportional reasoning. Proportional reasoning is foundational for algebra!

Real-world problems are laden with ratios, and leaning into this provides a foundation for student understanding. For example, a ratio may describe the relationship between two quantities: if there are 30 students in a class, and 20 of them are boys, the ratio of boys to girls is 20 boys to 10 girls, or 20:10, or 2:1.

It's also important to note that ratios can represent part-to-whole relationships as well as part-to-part relationships. All fractions are ratios, including the part-to-whole fractions students master in earlier grades. Most students master part-to-whole relationships when working with fractions in earlier grades and are later introduced to part-to-part relationships in middle school.

Rates describe relationships between two different units, usually with the word "per" in the middle. For example, if bananas cost $1 for 4 pounds, how much is the price per pound for those bananas? Other examples are price per gallon, inches per year, and cups per container.

In addition to this real-world application, the study of ratios and rates is crucial for interpreting data and transitioning to more advanced math concepts like algebra, including slope and functions. An understanding of ratios and rates is the bridge between basic arithmetic and more complex mathematics.

Typical Trajectory in Most State Standards Frameworks:

- Grades 1–5: Introduction to fractions as part-to-whole relationships (not yet identified as ratios)
- Grades 6+: Introduction of ratios as part-to-part relationships; introduction of rates

Ratios

Ratios compare two or more numbers and can be written in different ways.

Here are a few different ways to compare the balls in Christian's toy box.

Source: Ball images from istock.com/Nadezhda Kurbatova

You can use a ratio to compare a part to a whole.	The ratio of baseballs to all balls is 2 to 7. For every 2 baseballs in the box, there are 7 total balls. This ratio can be written as: $2:7$ or $\frac{2}{7}$
You can use a ratio to compare one part to another part.	The ratio of baseballs to beach balls is 2 to 1. For every 2 baseballs in the box, there is 1 beach ball. This ratio can be written as: $2:1$ or $\frac{2}{1}$
You can use a ratio to compare one whole group to another whole group.	The ratio of Christian's balls to his brother Mark's balls is 7 to 10. For every 7 balls Christian owns, Mark owns 10. This ratio can be written as: $7:10$ or $\frac{7}{10}$

Using Ratios as Rates

Rates are ratios that show two different units and how they relate to each other. When you have a rate, you can find an unlimited number of equivalent ratios (ratios that are equal). You can use ratios as rates to convert from one measurement unit to another, such as from inches to yards or from days to seconds.

Video 14: Ratios and Rates With Cuisenaire® Rods: You can use Cuisenaire® rods to help students solve problems using ratios and rates. *Pictured here: wooden Cuisenaire® rods*

https://qrs.ly/uag99lr

Converting Measurement Units

1. Write a ratio that has the unit you already know at the bottom and the unit you are converting to on the top. This ratio is called the **conversion factor**.

2. Then you multiply the **conversion factor** by the unit you already know.

How many feet of fencing must be purchased to go around a pool deck that has a perimeter of 24 yards?

Step 1

There are 3 feet in 1 yard, so the conversion factor is:

$$\frac{\text{feet}}{\text{yards}} \longrightarrow \frac{3}{1}$$

Step 2

I need to figure out how many feet there are in 24 yards, so I set up my ratio like this:

$$\frac{\text{feet}}{\text{yards}} \longrightarrow \frac{3}{1} \times 24 \text{ yards} = 72 \text{ feet}$$

[3 3 3 3 3 3 3 3 3 3 3 3 3 3 3 3 3 3 3 3 3 3 3 3]

$$24 \times 3 = 72$$

I need to purchase 72 feet of fencing to go around the pool deck.

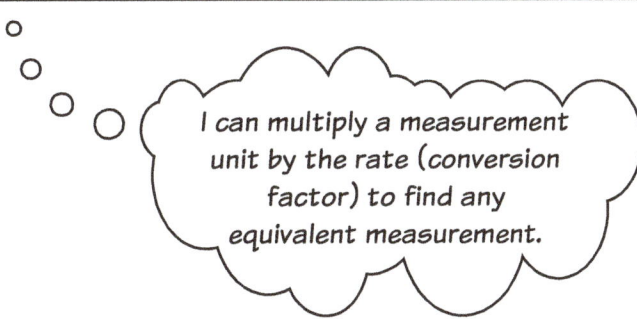

I can multiply a measurement unit by the rate (conversion factor) to find any equivalent measurement.

Equivalent Ratios

You can use several different strategies to illustrate **equivalent ratios**. Equivalent ratios are equal to one another.

Tape Diagram

Sam is making lemonade. The recipe calls for 3 parts of lemon juice for every 4 parts of water. If Sam has already poured 8 cups of water into his container, how much lemon juice should he add?

Lemon Juice: ⬛⬛⬛   Water: ⬛⬛⬛⬛
    3 parts   4 parts

Since 8 cups of water can be divided into 4 parts of 2 cups of water:

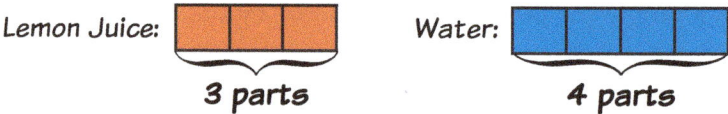

Then each part of lemon juice would also be 2 cups:

Lemon Juice: | 2c | 2c | 2c |

So, Sam needs to add 6 cups of lemon juice to make his lemonade.

Double Number Line

Molly is walking her dog. She walks 5 meters every 2 seconds. How far will she walk in 8 seconds?

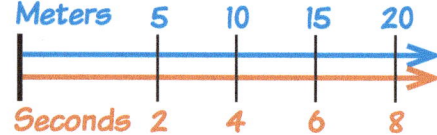

If Molly walks a constant speed, she will travel **20 meters** in **8 seconds**.

Equivalent Ratios (continued)

Ratio Table

A recipe calls for 2 ounces of pear juice to be added for every 5 ounces of apple juice. If I have 8 ounces of pear juice, how much apple juice should I add?

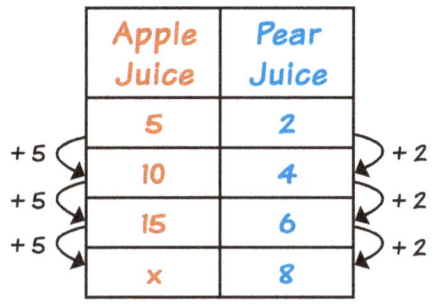

> I can repeatedly add 5 ounces of apple juice for every 2 ounces of pear juice and record it on a table.

15 + 5 = 20

I need 20 ounces of apple juice.

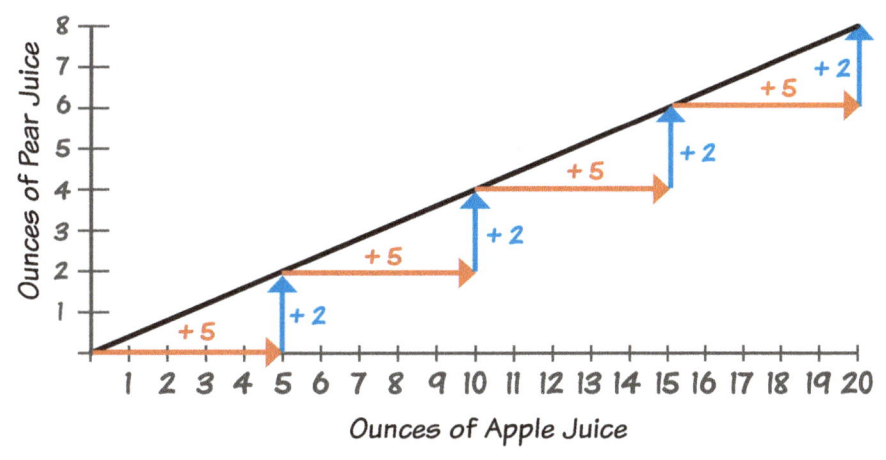

> I can also show repeatedly adding 5 ounces of apple juice for every 2 ounces of pear juice on a graph.

Equivalent Ratios (continued)

If 2 pounds of beans cost $5, how much will 9 pounds of beans cost?

Strategy 1
Rate Table

Pounds	$
1	2.50
2	5
4	10
6	15
8	20

20 + 2.50 = 22.50

If 8 pounds cost $20 and one more pound is $2.50, that makes $22.50 in all for 9 pounds of beans.

Strategy 2
Per-Unit Rates

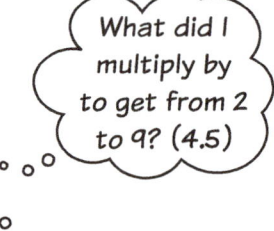

If 1 pound is half of the price of 2 pounds, that's $5 ÷ 2 or $2.50. I can multiply that by 9 to find the cost of 9 pounds.
9 × $2.50 = $22.50.

Strategy 3
Multiplicative Reasoning

① Set up a table to show the relationship between the amount of beans and the cost (c).

② Find the number that you multiply the "Rate We Know" to get to the "Rate We Want."

③ Multiply the other "Rate We Know" by the result from Step 2 to find the missing value (c).

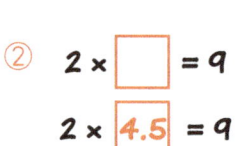

①	Rate We Know	Rate We Want
Pounds	2 ×4.5→	9
$	5 ×4.5→	c

② 2 × ☐ = 9

2 × 4.5 = 9

③ 5 × 4.5 = c

c = 22.50

What did I multiply by to get from 2 to 9? (4.5)

I can multiply 5 by the same number (4.5) to find how much 5 pounds of beans will cost.

Chapter 12

Algebraic Expressions, Equations, and Inequalities

Algebraic thinking begins in primary grades when students work with missing numbers in equations and word problems. The missing numbers are typically represented with an open box, a space, or a symbol such as a question mark or a star. Examples include $11 + \square = 16$, $11 + ? = 16$, and $11 + \underline{} = 16$.

Young children also encounter algebraic thinking with context problems. Just as with the contextless equations earlier, a context problem might be represented with an equation that has a box, a space, or a symbol that represents an unknown number. For example: Calvin has 11 toy cars. Sam gave him some more. Now Calvin has 16 toy cars. How many cars was Calvin given? This context problem would be represented as $11 + \square = 16$.

Inequalities provide another way in which students think algebraically. In the early years, children understand and explain simple inequalities such as $5 + 4 \neq 10$, or $5 + 4 < 10$. These comparisons allow students to think abstractly, and verbalizing their thinking is an important step. Working with equations and inequalities in the early years lays a foundation for more formalized algebra in later years.

In most state standards frameworks, working with letters as variables begins somewhere around third grade. Although students are not yet using formalized methods for balancing equations (e.g., doing the same thing to both sides), using a letter to represent a missing value provides yet another foundation for later.

What we recognize as "algebra" begins in the middle school years and continues into high school. These experiences provide students with opportunities to use more formalized methods to work with variables, balance equations, explore functions, graph linear relationships, and the like.

At both stages, early years and middle school, students are using algebraic reasoning to make sense of both context problems and mathematical problems. Furthermore, they begin to make generalizations. For example, with the introduction to algebra, students move away from finding the answer to a single problem and closer to knowing how to solve any problem like it. Students learn to make sense of mathematics, instead of making sense of only one problem. This is the power of algebra!

Typical Trajectory in Most State Standards Frameworks:

- Grades K–5: Find the missing value in context problems and mathematical problems
- Grade 1: Introduce "=" as a comparison that means "balance"
- Grade 1: Introduce $\neq$, $<$, and $>$ as comparison symbols
- Grade 3: Introduce letters as variables
- Grade 6+: Find the missing value in an equation using formalized methods

Algebra Vocabulary

Term
The parts of an expression separated by + or − signs, unless the signs are within parentheses.

$6x^2 - 5x + (4 - 2x) + 2$

This expression has four **terms**: $6x^2$, $5x$, $(4 - 2x)$ and 2.

Sum
The result of an addition problem.

$a + b = ⓒ$

↙ Sum

Product
The result of a multiplication problem.

$18 \cdot c = ⓘⓑⓒ$

↙ Product

Difference
The result of a subtraction problem.

$m - j = ⓚ$

↙ Difference

Quotient
The result of a division problem.

$\dfrac{14x}{2} = ⓨ$

↙ Quotient

Coefficient
A number used to multiply a variable.

④ $x - 7 = 5$

Coefficients

⑥z

Factor
An integer that divides evenly into another.

$2 \times 6 = 12$

2 and 6 are **factors** of 12.

Variable
A symbol (any letter or ☐) that stands for a number you don't know yet.

$n + 2 = 6$

$7 + 9 = ☐$

n and ☐ are **variables**.

$n + 2 = 7$
☐ $+ 2 = 7$

 Both letters and boxes can be used to represent variables.

Writing Expressions

An **expression** is a group of terms that represent numbers, unknowns, and operations. An expression does not have an equal sign. When you need to translate a word problem into an expression, you use numbers when you know what they are. You use variables when you don't know the numbers.

Problem	Expression
Show a full bag of oranges plus 3 more oranges.	$b + 3$
Show a box of apples with 3 missing.	$b - 3$
Show 3 full bags of kiwi fruits.	$3b$ or $3 \times b$
Show a full box of pomegranates separated into groups of 3.	$b \div 3$ or $\frac{b}{3}$

Problem	Expression
Add 8 and 7, then multiply by 2.	$2(8 + 7)$
Add 8 and 27, then add 2 more.	$(8 + 27) + 2$
Multiply 6 by 30 and add it to the product of 6 times 7.	$(6 \times 30) + (6 \times 7)$

Sometimes it is not necessary to calculate an answer but instead to be able to recognize the comparative size of the answer.

- $5(12 + 9)$ is 5 times larger than $12 + 9$.
- $3(18{,}972 + 921)$ is three times larger than $18{,}972 + 921$.

Algebraic Expressions, Equations, and Inequalities

Evaluating Expressions

When you **evaluate** an **expression**, you substitute the given number for every variable you find in an expression.

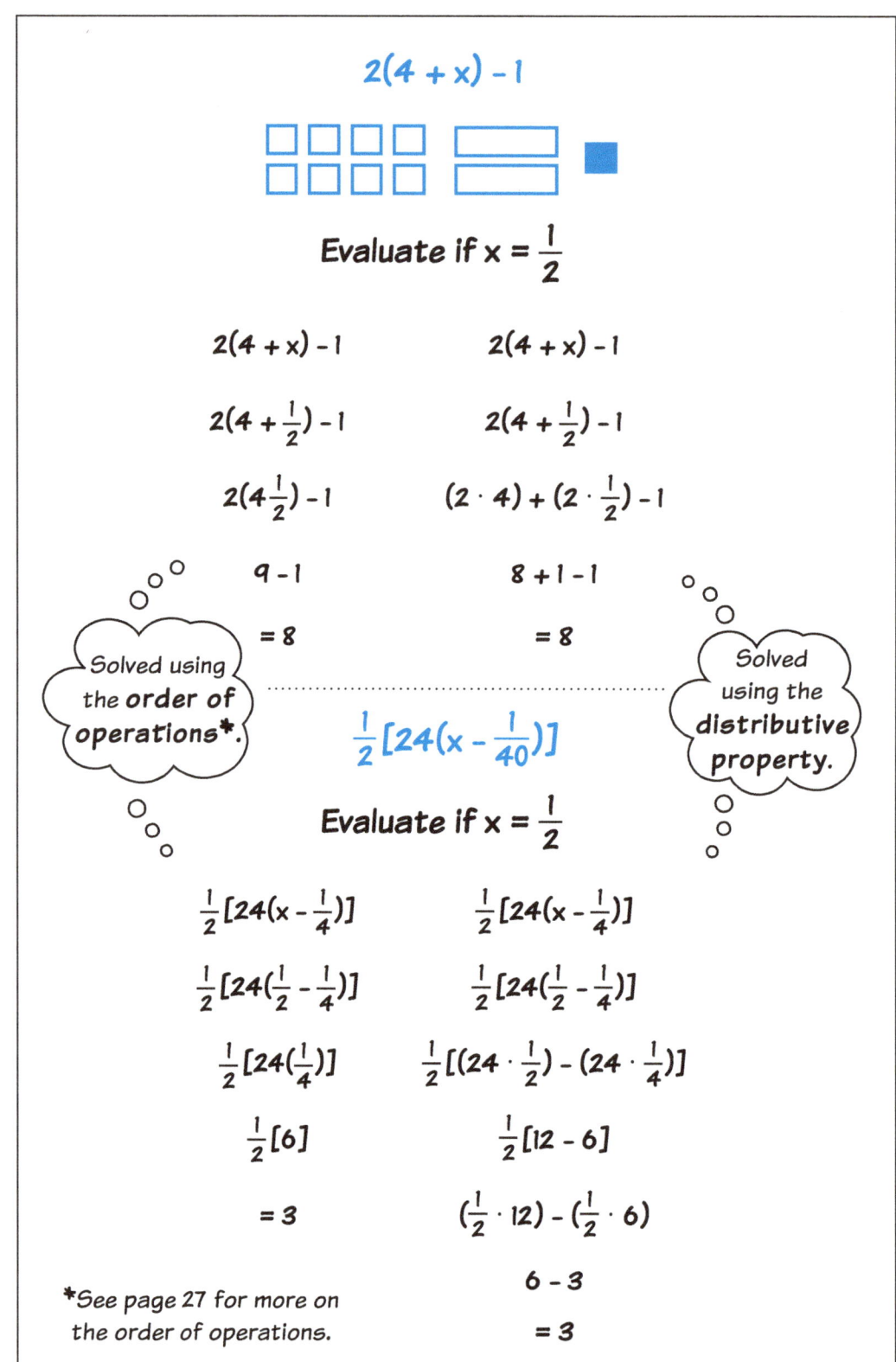

$2(4 + x) - 1$

Evaluate if $x = \frac{1}{2}$

$2(4 + x) - 1$	$2(4 + x) - 1$
$2(4 + \frac{1}{2}) - 1$	$2(4 + \frac{1}{2}) - 1$
$2(4\frac{1}{2}) - 1$	$(2 \cdot 4) + (2 \cdot \frac{1}{2}) - 1$
$9 - 1$	$8 + 1 - 1$
$= 8$	$= 8$

*Solved using the **order of operations**.*   *Solved using the **distributive property**.*

$\frac{1}{2}[24(x - \frac{1}{4})]$

Evaluate if $x = \frac{1}{2}$

$\frac{1}{2}[24(x - \frac{1}{4})]$	$\frac{1}{2}[24(x - \frac{1}{4})]$
$\frac{1}{2}[24(\frac{1}{2} - \frac{1}{4})]$	$\frac{1}{2}[24(\frac{1}{2} - \frac{1}{4})]$
$\frac{1}{2}[24(\frac{1}{4})]$	$\frac{1}{2}[(24 \cdot \frac{1}{2}) - (24 \cdot \frac{1}{4})]$
$\frac{1}{2}[6]$	$\frac{1}{2}[12 - 6]$
$= 3$	$(\frac{1}{2} \cdot 12) - (\frac{1}{2} \cdot 6)$
	$6 - 3$
	$= 3$

*See page 27 for more on the order of operations.

Generating and Analyzing Patterns

A **pattern** is a sequence that repeats the same process over and over. A **T-Chart** is a tool used to help see number patterns.

There are 2 marbles in the jar. Each day, 4 marbles are added. How many marbles are in the jar for each of the first 5 days?

Day	Marbles
0	2
1	6
2	10
3	14
4	18
5	22

+4 between each day.

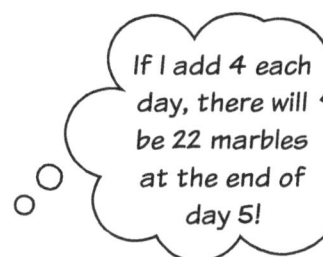

If I add 4 each day, there will be 22 marbles at the end of day 5!

Investigating number patterns leads to identifying rules and observing unique features.

Pattern	Rule	Features
5, 10, 15, 20, …	start with 5, add 5	The numbers are multiples of 5. They all end in a 5 or 0. The numbers that end in a 0 are products of 5 and an even number.
3, 8, 13, 18, 23, …	start with 3, add 5	The numbers alternatively end with a 3 or 8.

Algebraic Expressions, Equations, and Inequalities

Writing Equations..........................

An **equation** is a mathematical sentence. It always has an = sign showing that the expressions are equal. When you need to translate a word problem into an equation, think about the quantities that are equal to each other. Then write an expression for each quantity.

Susan has 15 pieces of gum. How many pieces are in each pack if she has 2 packs and 3 extra pieces?

$15 = 2p + 3$

p is the number of pieces of gum in each pack.

Sam gets $5 per hour for doing yard work, plus $20 when he finishes the yard. If he made $35, how many hours did he work?

$5h + 20 = 35$

h is the number of hours spent doing yard work.

Daniel went to visit his grandmother, who gave him $5.50. Then he bought a book costing $9.20. If he has $2.30 left, how much money did he have before visiting his grandmother?

Expression $x + 5.50 - 9.20$
Equation $x + 5.50 - 9.20 = 2.30$

x is the amount of money Daniel had before visiting his grandmother.

How many 44¢ stamps can you buy with $11?

$11 \div 0.44 = n$

or

$0.44n = 11$

These are both equations!

n is the number of stamps you can buy.

Algebraic Expressions

Solving Equations (Two-step)

When you **solve** an **equation**, you find the values for the variables that make it true.

 Whatever you do to one side of an equation, you must also do to the other side!

 Addition and subtraction "undo" each other. They are called **inverse operations.** Multiplication and division "undo" each other, also.

Video 15: Ten Frames and Counters: You can use blank ten frames as variables and counters as constants to help students "see" how to solve an equation. *Pictured here: KP® Ten-Frame Tiles*

https://qrs.ly/ydg99m0

Video 16: Algebra Tiles: You can use algebra tiles to help students "see" how to solve an equation. *Pictured here: KP® Ten-Frame Tiles*

https://qrs.ly/a2g99m4

$$15 = 2p + 3$$

① Subtract 3 from each side of the equation.

② Divide both sides by 2, so that you can find the value of 1 p.

① $15 - 3 = 2p + 3 - 3$
$12 = 2p$

② $\dfrac{12}{2} = \dfrac{2p}{2}$
$6 = p$

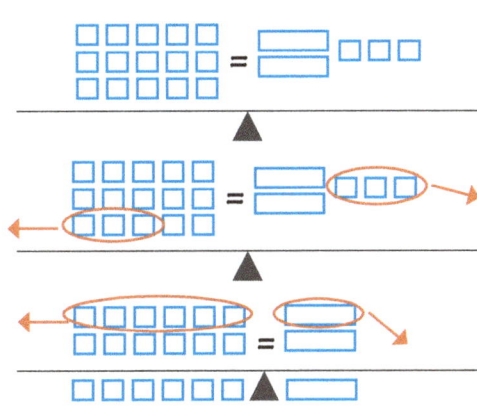

$$5h + 20 = 35$$

① Subtract 20 from each side of the equation.

② Divide both sides by 5, so that you can find the value of 1h.

① $5h + 20 - 20 = 35 - 20$
$5h = 15$

② $\dfrac{5h}{5} = \dfrac{15}{5}$
$h = 3$

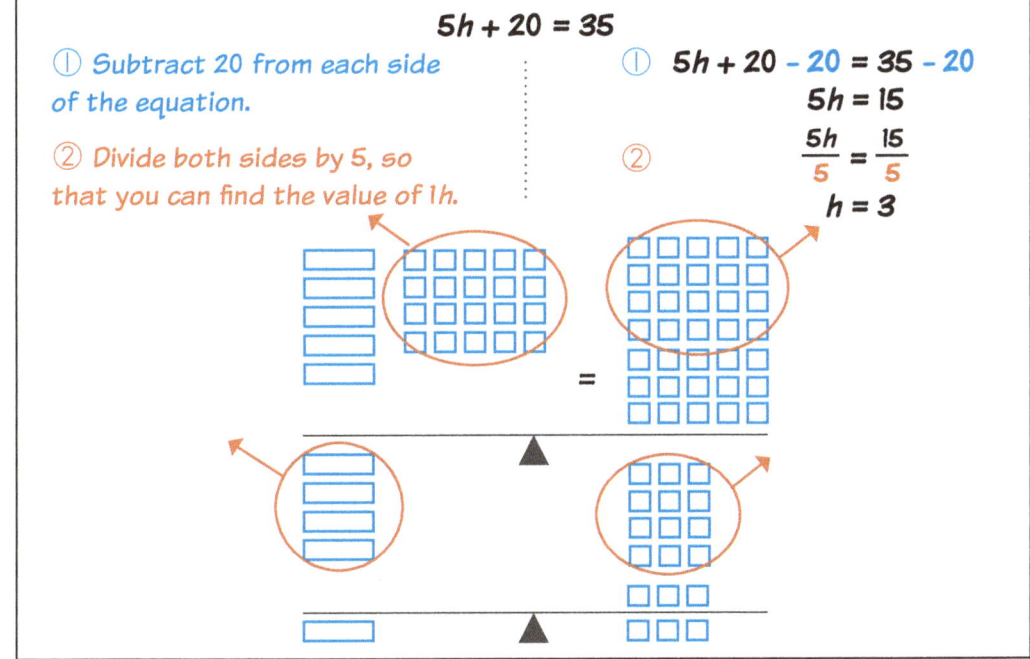

Algebraic Expressions, Equations, and Inequalities

Solving Equations (continued)

Many real world problems involve two **variables** that change in relationship to each other.

> A car is traveling on a highway at a speed of 60 miles per hour. Express this constant rate of speed in relationship to distance and time. Then determine the distance the car has traveled after 1, 2, and 3 hours of traveling.
>
> $$d \;=\; 60 \;\cdot\; t$$
>
> distance in miles (dependent variable)   miles per hour   time in minutes (independent variable)
>
> | Time (hours) | Distance (miles) |
> |---|---|
> | 1 | 60 |
> | 2 | 120 |
> | 3 | 180 |
>
>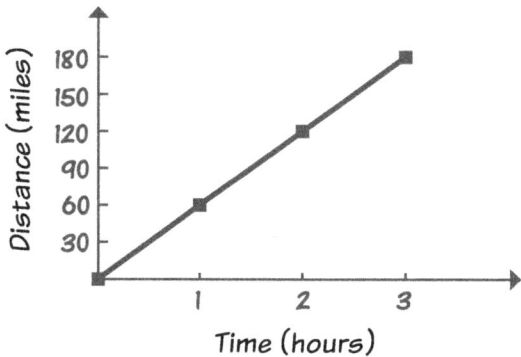
>
> A **variable** is anything you are trying to measure. There are two types of variables—independent and dependent.
>
> An **independent variable** stands alone and isn't changed by the other variables you are trying to measure. In fact, it causes a change in the other variables or dependent variables.
>
> A **dependent variable** is something that depends on other factors.
>
> To decide which variable is independent and which is dependent, just insert the names into this sentence:
>
> <u>Independent variable</u> causes a change in <u>dependent variable</u>, and it is not possible that <u>dependent variable</u> could cause a change in <u>independent variable</u>.
>
> For example: <u>Time</u> causes a change in <u>distance</u>, and it isn't possible that <u>distance</u> could cause a change in <u>time</u>.

Solving Inequalities

While an equation has an = sign to show that the expressions are equal, an **inequality** does not contain the = sign. Instead, an inequality contains a different sign, like those shown below:

$\neq$	$>$	$<$	$\geq$	$\leq$
not equal	greater than	less than	greater than or equal	less than or equal

 You solve **inequalities** with + and − in the same way that you solve equations: whatever you do to one side of an inequality, you must also do to the other side.

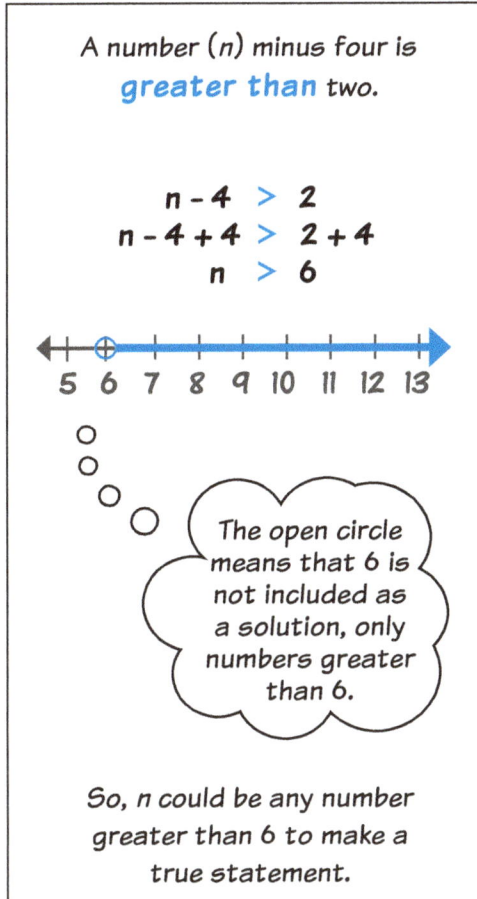

A number (n) minus four is **greater than** two.

$$n - 4 > 2$$
$$n - 4 + 4 > 2 + 4$$
$$n > 6$$

The open circle means that 6 is not included as a solution, only numbers greater than 6.

So, n could be any number greater than 6 to make a true statement.

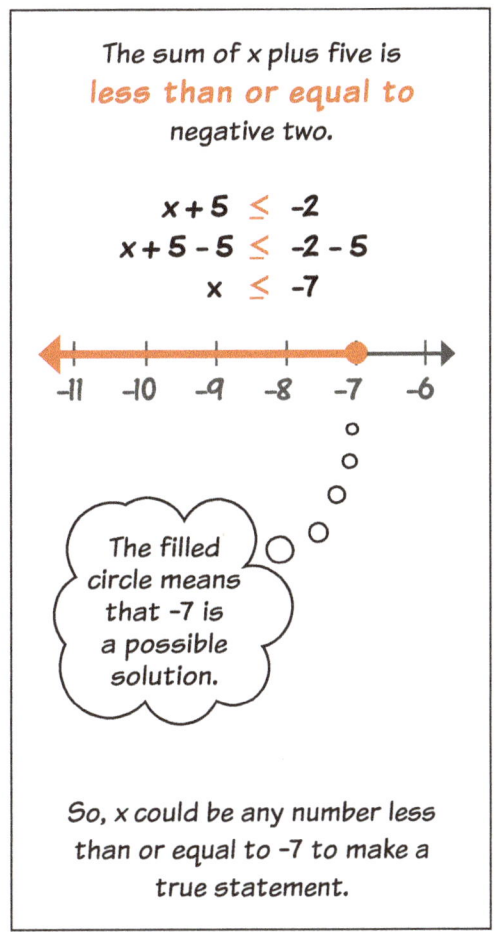

The sum of x plus five is **less than or equal to** negative two.

$$x + 5 \leq -2$$
$$x + 5 - 5 \leq -2 - 5$$
$$x \leq -7$$

The filled circle means that −7 is a possible solution.

So, x could be any number less than or equal to −7 to make a true statement.

 Inequalities have an endless amount of possible solutions, so it's easier to show the answers on a number line.

Algebraic Expressions, Equations, and Inequalities

Chapter 13

Coordinate Planes

Coordinate planes help us describe where an object is located in space. For example, in real life, coordinate planes can conveniently be found when maps are printed on a grid and used to identify the location of a street or store. In math class, coordinate planes typically represent spatial relationships between points, lines, and shapes.

Typically during fifth grade, students focus exclusively on the first quadrant. Beginning in sixth grade, students work with negative integers for the first time, providing the foundation needed to work in all four quadrants.

Also in the later grades, coordinate planes provide a way for students to represent the relationship between variables. We often refer to this as "graphing the equation," "graphing the function," etc. This, in turn, allows them to move past plotting points, now graphing lines and analyzing trends. These skills provide a crucial foundation for understanding and solving problems not only in math but also in fields such as science and engineering.

Typical Trajectory in Most State Standards Frameworks:

- Grade 5: Students graph points in quadrant 1 of the coordinate plane
- Grades 6+: Students graph points in all four quadrants of the coordinate plane

Coordinate Geometry

A **coordinate grid** is a way to locate points on a flat surface. To draw a coordinate grid, you draw a horizontal line (called the *x-axis*) and a vertical line (called the *y-axis*). The point where these two lines intersect is called the *origin*.

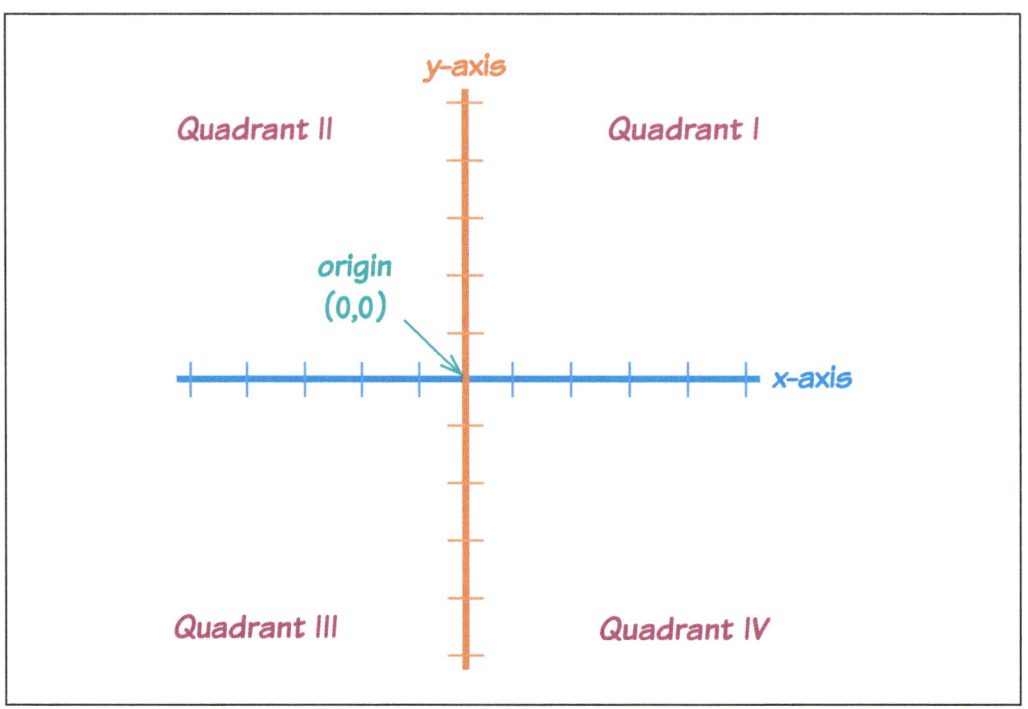

The axes divide the plane into 4 sections called **quadrants**.

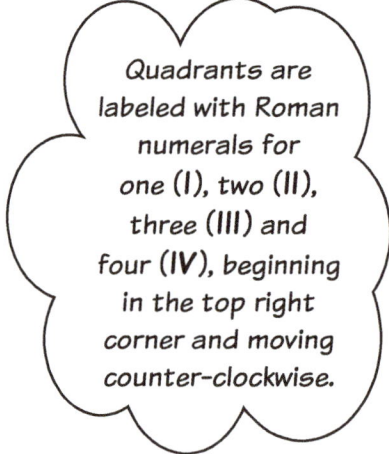

Quadrants are labeled with Roman numerals for one (I), two (II), three (III) and four (IV), beginning in the top right corner and moving counter-clockwise.

Coordinate Plane..........................

You can name any point on a **coordinate plane** with two numbers (called **ordered pairs** or coordinates). The first number shows how far the point is side-to-side along the **x-axis**, and is called the **x-coordinate**. The second number is for how far the point is up-and-down along the **y-axis**, and is called the **y-coordinate**.

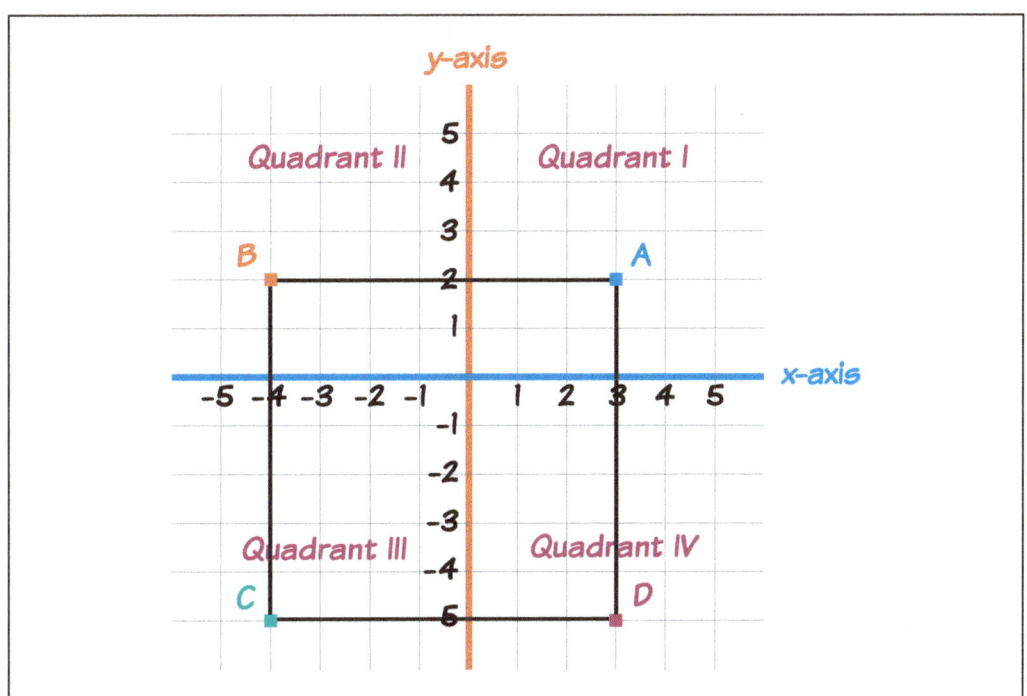

For rectangle ABCD:

Point	Ordered Pair	Quadrant
A	(3, 2)	Quadrant I
B	(-4, 2)	Quadrant II
C	(-4, -5)	Quadrant III
D	(3, -5)	Quadrant IV

To find the length of $\overline{AB}$, you can add the absolute value* of each x-coordinate: $|3| + |-4| = 3 + 4 = 7$.

To find the length of $\overline{AD}$, you can add the absolute value* of each y-coordinate: $|2| + |-5| = 2 + 5 = 7$.

*See page 92 for more on absolute value.

Coordinate Planes

145

Graphing in the Coordinate Plane ....

When graphing points on a **coordinate plane**, the value of y depends on the value of x. This means that we can choose a value for x, but as soon as we do, the value of y changes.

 x = independent variable

y = dependent variable

$y = 3(x) + 2$

If we choose the x values 1, 2, 3, and 4, what will the y values be?

x	Substitute x into the equation to determine y	y
1	3(1) + 2 = 3 + 2 = 5	5
2	3(2) + 2 = 6 + 2 = 8	8
3	3(3) + 2 = 9 + 2 = 11	11
4	3(4) + 2 = 12 + 2 = 14	14

The ordered pairs for this function would be:

(1, 5), (2, 8), (3, 11), (4, 14)

x	y	(x, y)
1	5	(1, 5)
2	8	(2, 8)
3	11	(3, 11)
4	14	(4, 14)

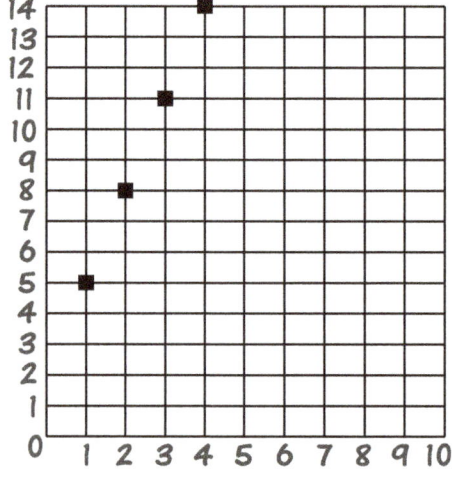

Matt ate lunch with his cousin, who lived 2 miles from his house. He then continued in the same direction as he walked to his grandma's house at a rate of 3 mph. After walking 1 hour, how far was Matt from his house?

I can use equations, tables, and graphs to model and solve real-world problems.

146 SEEING THE MATH YOU TEACH, GRADES K–6

Graphing Numerical Patterns

Coordinate grids can also be used to analyze two number patterns generated by using given rules. There are several things we need to know about patterns: the rule, the pattern, and the features of the pattern.

Rule	Pattern	Features
start with 0, add 3	0, 3, 6, 9, 12, 15, 18, 21,...	The sum of the digits is a multiple of 3.
start with 0, add 6	0, 6, 12, 18, 24, 30, 36,...	All of the numbers are even. All of the numbers are multiples of 6.

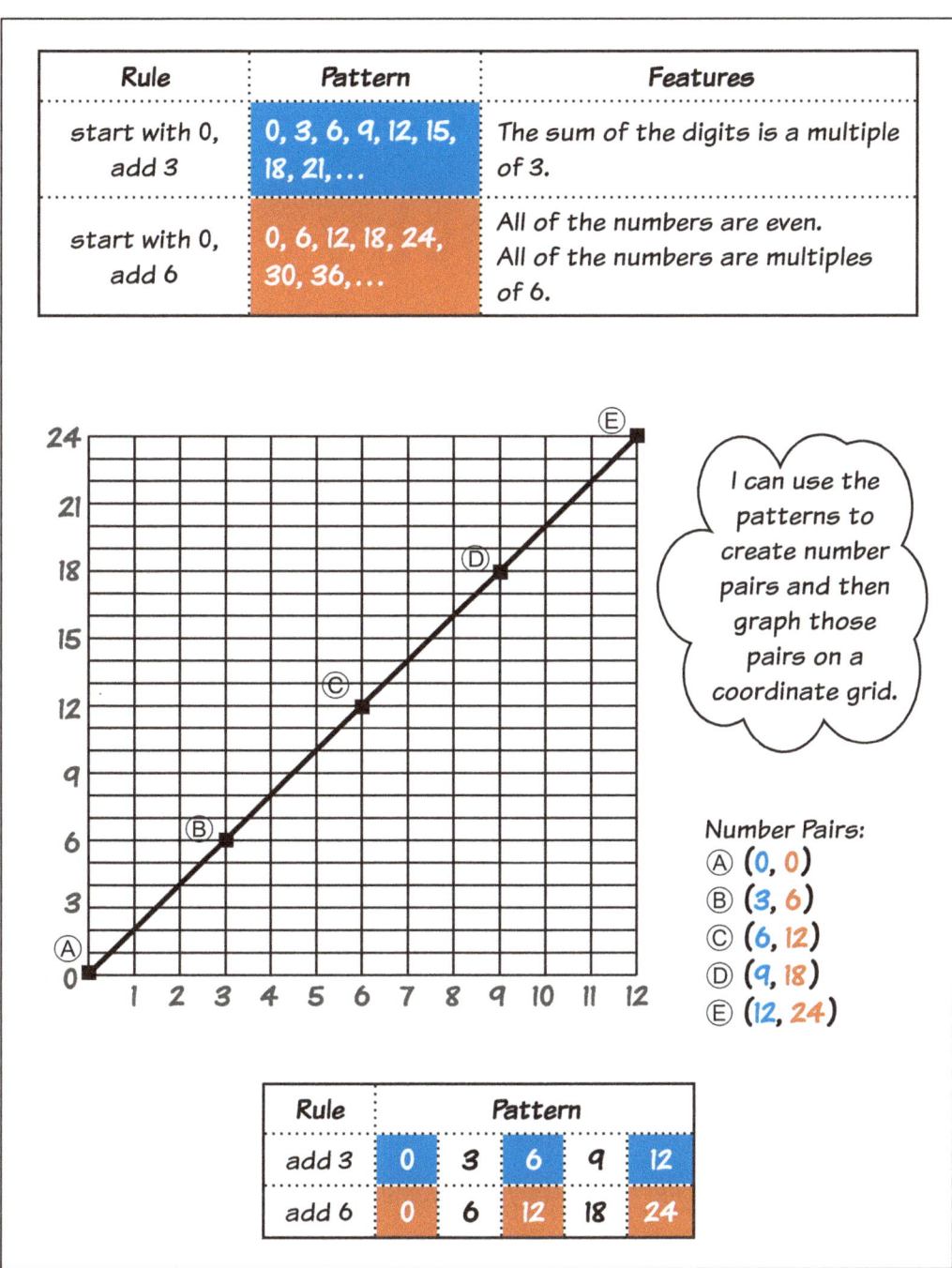

I can use the patterns to create number pairs and then graph those pairs on a coordinate grid.

Number Pairs:
- Ⓐ (0, 0)
- Ⓑ (3, 6)
- Ⓒ (6, 12)
- Ⓓ (9, 18)
- Ⓔ (12, 24)

Rule	Pattern				
add 3	0	3	6	9	12
add 6	0	6	12	18	24

Coordinate Planes

Chapter 14

Geometry

In Grades K–6, geometry focuses on properties and relationships between points, lines, shapes, surfaces, and other attributes of two- and three-dimensional figures. Geometry is a foundation for spatial reasoning, logical thinking, and application in fields like science, engineering, visual arts, and so many other fields. We often use geometric thinking when playing games and working on building projects, as well.

When students study geometry, they are doing more than merely learning the names and attributes of shapes. They are learning to "see" shapes in their mind's eye—mentally sliding, turning, and flipping them to see them from different vantage points. They are using spatial relationships to solve problems and connect to measurement concepts such as area, perimeter, and volume.

In the early years, young students identify shapes based on their appearance and overall features, often using nonmathematical language to describe them. They put shapes together to build new shapes (compose) or break shapes apart into other shapes (decompose). This connects to the number work they're doing during those years, as well—connecting to the idea, fundamental to the study of algebra, that both whole numbers and fractions can be composed and decomposed.

After much exploration with two- and three-dimensional shapes, students continue working with shapes and figures, identifying and naming the properties of each, such as the number of sides of a polygon. At this stage, they don't necessarily see relationships between those properties, yet that will change.

Eventually, students will begin to reason logically about geometric properties, classify shapes and figures based on their characteristics, and understand simple relationships between different shapes.

It's important to know that as students go through these developmental stages, the sophistication of their understanding is based on experience, not age. Therefore, students need many, many opportunities to develop their spatial skills at all times, not just when the math program sets aside a few weeks for the study of geometry.

Typical Trajectory in Most State Standards Frameworks:

- Grades K–2: Name, identify, build, compose, and decompose 2D and 3D shapes; identify specified attributes of 2D and 3D figures
- Grades 3–5: Categorize and classify shapes by attributes; identify lines of symmetry
- Grade 6: Represent 3D figures using nets composed of 2D shapes

Attributes of Shapes

Shapes are most often described or categorized by their **defining attributes**.

Shape Attributes

- Number of sides
- Number of angles
- Vertex/vertices (pages 157, 161)
- Congruent sides (page 152)
- Adjacent sides (page 152)
- Parallel sides (page 151)
- Perpendicular sides (page 151)
- Symmetry (page 152)
- Angle types (pages 153–154)

Defining Attribute

These shapes are triangles. They each have 3 sides and **3 vertices** (angles/corners).

A **defining attribute** is a specific feature or characteristic such as the number of sides, the number of vertices, or the length of each side.

Non-defining Attribute

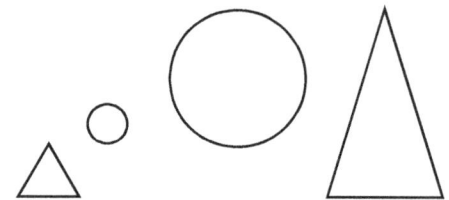

The shapes on the right are big, and the shapes on the left are small.

A **non-defining attribute** is a characteristic that could describe a variety of shapes, such as color or size.

Geometry Vocabulary

Point

An exact position in space.

This is point K.

Line

A straight path of points that has no endpoints.

This is line FG.

Ray

A line that has 1 endpoint and continues indefinitely in the other direction.

This is ray OP.

Line Segment

Part of a line that has 2 endpoints.

This is line segment SF.

Parallel

Parallel lines never cross because, like rails on a train track, they are always the same distance apart no matter how far the lines extend.

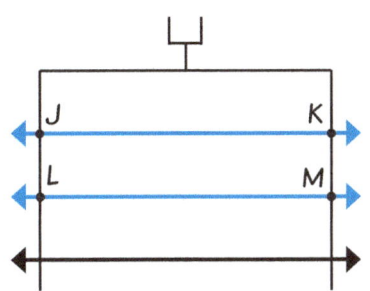

The yard lines on a football field are **parallel**.

$\overleftrightarrow{JK} \parallel \overleftrightarrow{LM}$

Line $\overleftrightarrow{JK}$ is **parallel** to line $\overleftrightarrow{LM}$.

Perpendicular

Perpendicular lines form a right angle (90°) where they intersect.

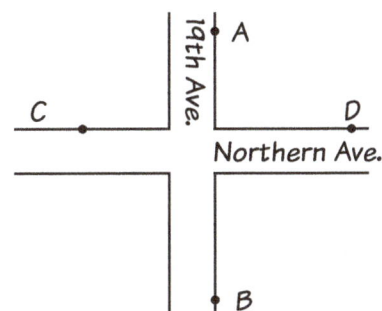

Northern Avenue and 19th Avenue are **perpendicular** roads.

$\overleftrightarrow{AB} \perp \overleftrightarrow{CD}$

Line $\overleftrightarrow{AB}$ is **perpendicular** to line $\overleftrightarrow{CD}$.

Symmetry

When you can fold a figure so that it has two parts that match exactly, it is said to have a **line of symmetry**.

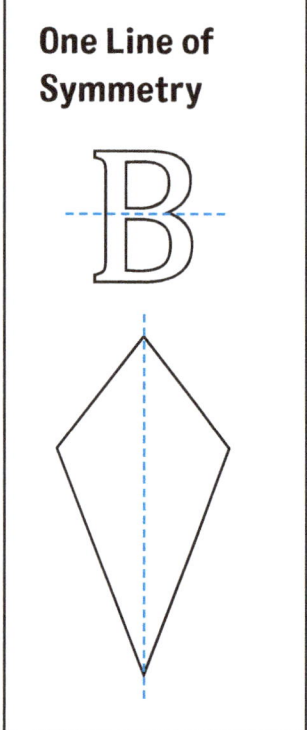

One Line of Symmetry

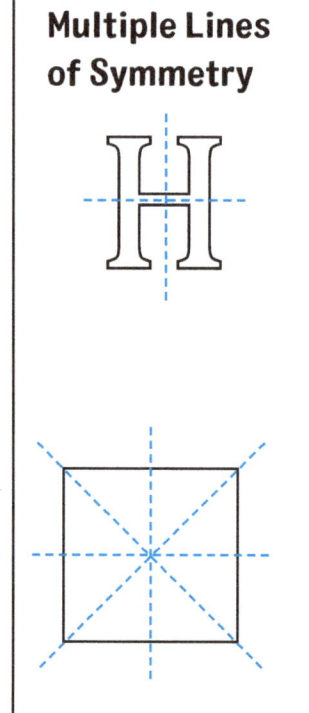

Multiple Lines of Symmetry

No Lines of Symmetry

Adjacent and Congruent Sides

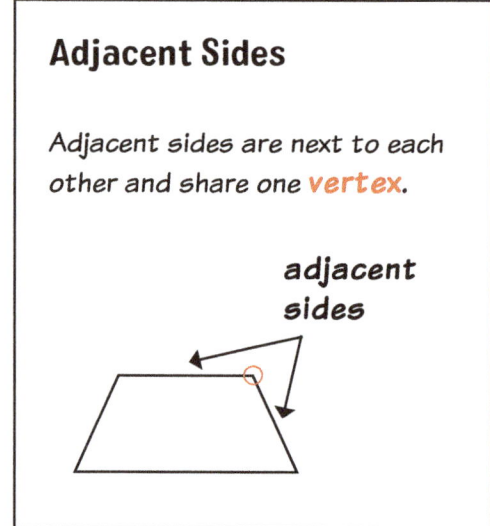

Adjacent Sides

Adjacent sides are next to each other and share one **vertex**.

adjacent sides

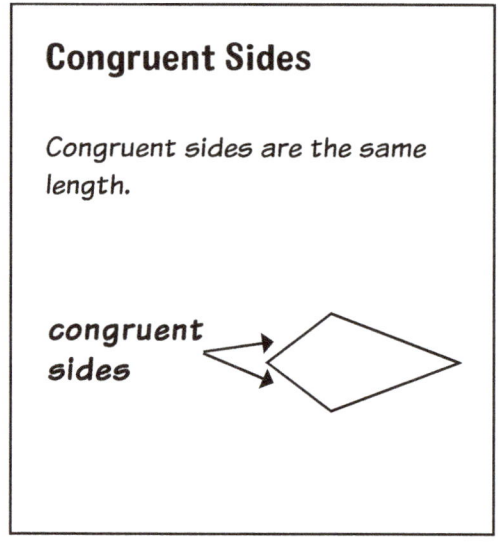

Congruent Sides

Congruent sides are the same length.

congruent sides

Angles

An **angle** is a turn around a point. The size of an angle is determined by measuring how far one side is turned from the other side.

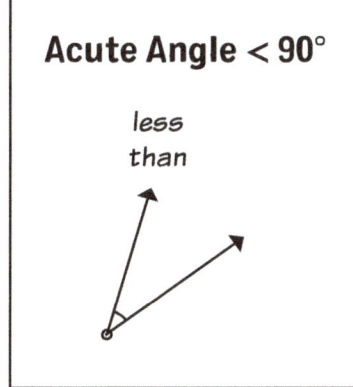
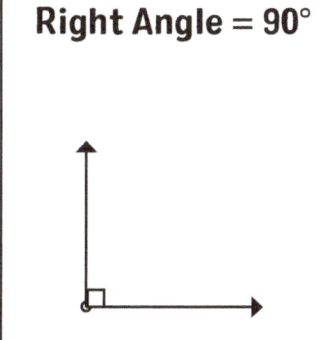
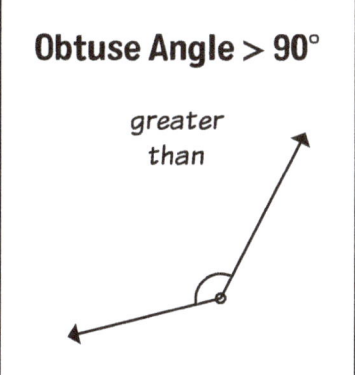

For more information about angle measures, see pages 154–155.

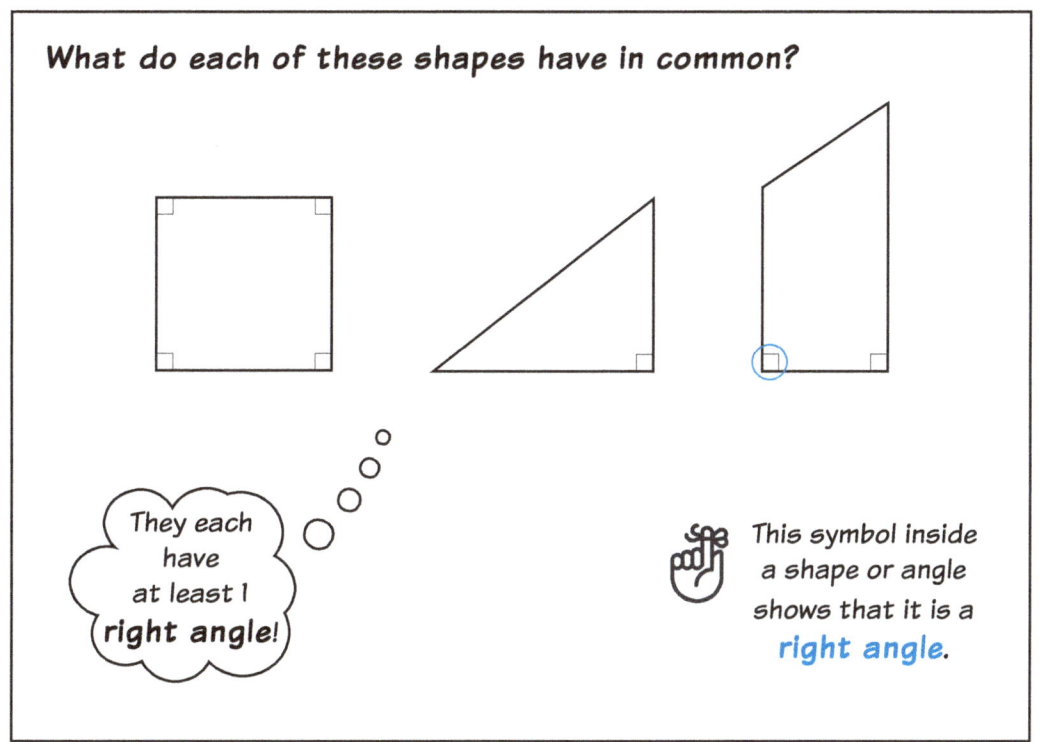

What do each of these shapes have in common?

They each have at least 1 **right angle!**

This symbol inside a shape or angle shows that it is a **right angle**.

Geometry 153

Angle Measures

An **angle** is formed when two rays or line segments meet at a common point, sometimes called a **vertex**. An angle is measured in reference to a circle with its center at the common endpoint of the rays. The measure of an angle tells how far one ray of the angle is turned from the other ray. It is measured in **degrees**. One full turn is 360°.

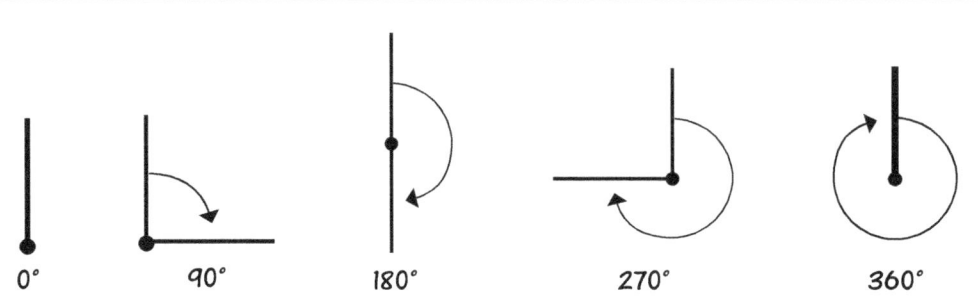

- 0° is no turn.
- 90° is a quarter turn.
- 180° is a half turn.
- 270° is a three-quarter turn.
- 360° is a full turn.

- **Acute angles** ⟶ less than 90°
- **Right angles** ⟶ exactly 90°
- **Obtuse angles** ⟶ between 90° and 180°
- **Straight angles** ⟶ exactly 180°

 An angle that is 1° is $\frac{1}{360}$ of a full circle.

An angle that is 45° is $\frac{45}{360}$ or $\frac{1}{8}$ of a full circle.

Measuring Angles

Unknown angles can be measured or determined by looking carefully at the measure of known angles.

Video 17: Angle Measurement With Paper Circles: You can use paper circles to help students "see" angles as "wedges" that fit between the two rays/ line segments of an angle. *Pictured here: paper circle*

https://qrs.ly/zag99m7

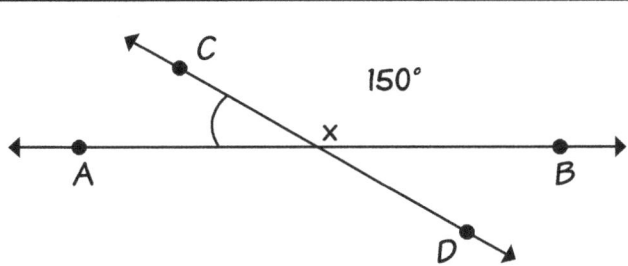

$\overleftrightarrow{AB}$ is a straight line, so ∠AXB is 180°.

I can calculate the measure of ∠AXC by subtracting 150° from 180°.

∠AXC + ∠CXB = 180°
∠AXC + 150° = 180°
∠AXC = 30°

See pages 153–154 for more about angles.

Angles can also be measured by using a **protractor**. The figure below shows a protractor being used to measure a 45° angle.

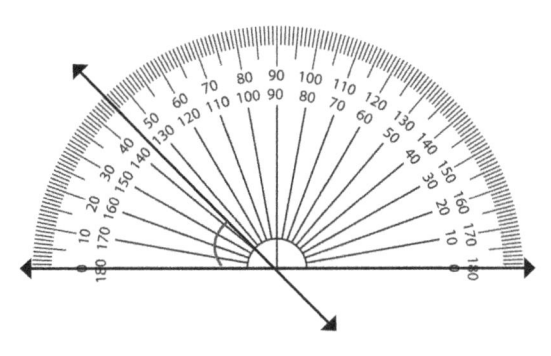

The protractor is placed so that one side of the angle lies on the line corresponding to 0°.

The measurement of the angle is read by noting where the other side of the angle passes through the protractor.

Source: Protracter image from iStock.com/Glam-Y

 Two numbers are shown on the protractor at each 10° increment, so students must use their knowledge of acute and obtuse angles—and of angle markings—to select the correct measurement.

Geometry

Polygons

A **polygon** is a closed figure formed by straight line segments.

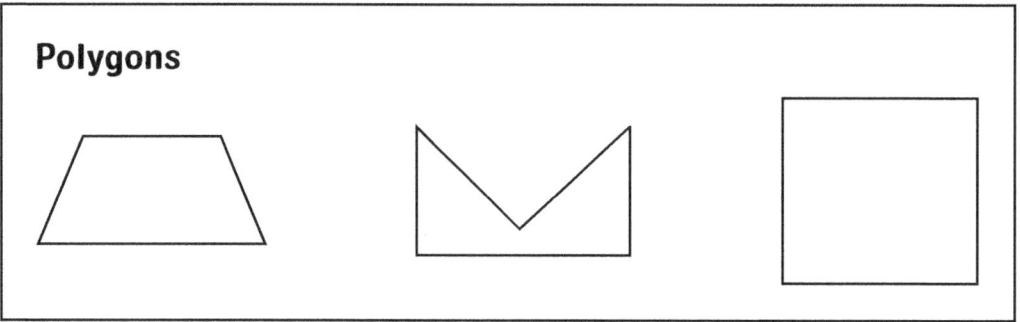

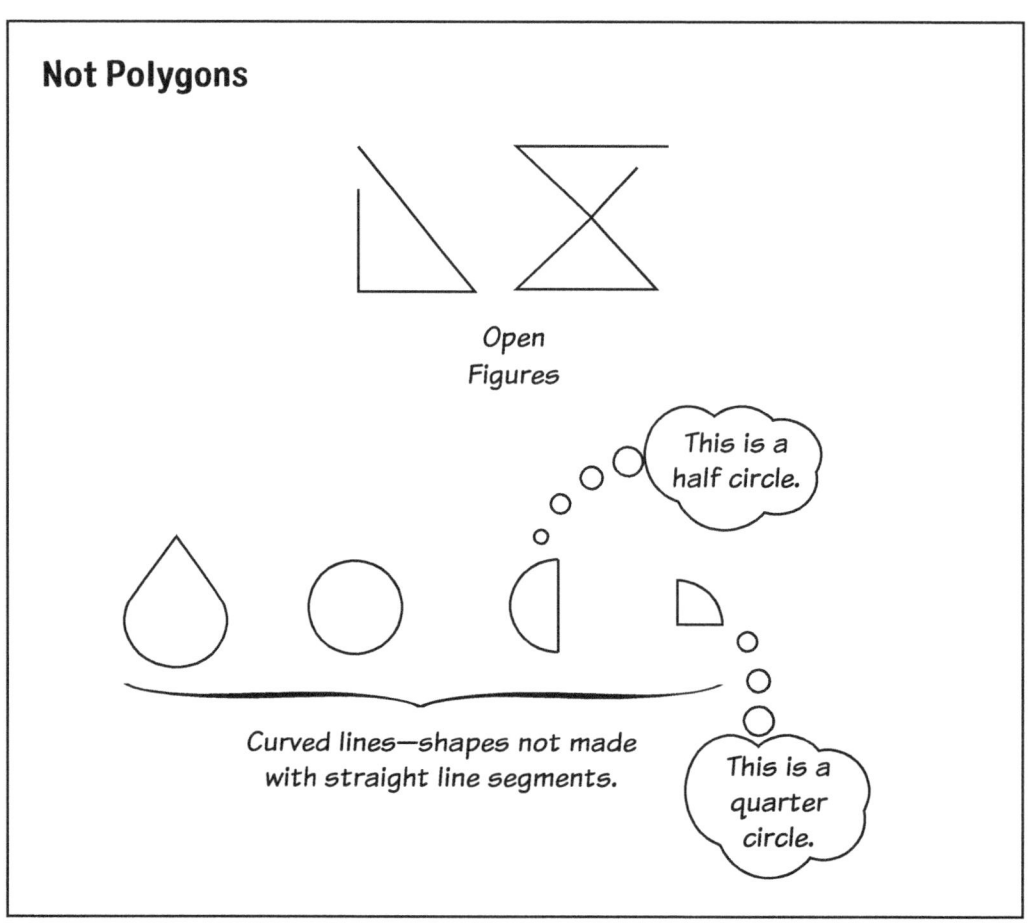

2-D Shapes

Shapes can be classified as either two-dimensional (**2-D**) or three-dimensional (**3-D**). 2-D shapes have length and width—they are considered "flat."

Regular polygons are shapes with equal sides and equal angles. Irregular polygons have sides that are not equal and angles that differ from one another.

Name	Examples	Description
Triangle		3 sides, 3 angles
Quadrilateral (4 sides, 4 angles)	Square	4 congruent sides, 4 equal angles
	Rectangle	opposite sides are parallel, 4 equal angles
	Trapezoid	at least 1 pair of parallel sides
	Parallelogram	2 pairs of parallel sides
	Rhombus	parallelogram with 4 equal sides
	Kite	2 pairs of adjacent, congruent sides
Pentagon		5 sides, 5 angles
Hexagon		6 sides, 6 angles
Octagon		8 angles, 8 sides
Decagon		10 angles, 10 sides

Rectangle

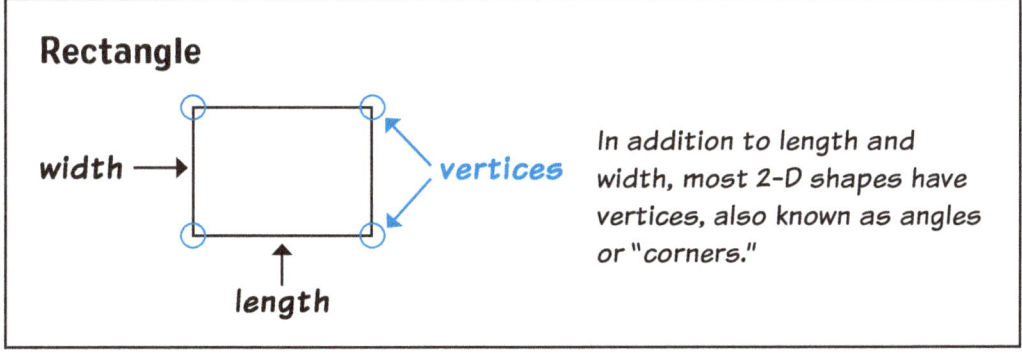

In addition to length and width, most 2-D shapes have vertices, also known as angles or "corners."

Geometry

Classification of Triangles

Triangles can be classified by either their sides or their **angles**.

Classification by Sides

 An **equilateral triangle** has 3 equal sides.

 An **isosceles triangle** has 2 equal sides.

 A **scalene triangle** has no equal sides.

Classification by Angles

 An **acute triangle** has 3 angles less than 90°.

 An **obtuse triangle** has 1 angle more than 90°.

 A **right triangle** has one 90° angle.

Classification of Quadrilaterals

Sometimes shapes can fit into more than one category.

This shape is a square.
It has four equal sides and four 90° angles. This shape is a special rectangle. It has opposite sides that are parallel and four 90° angles.

 All squares are rectangles, but not all rectangles are squares.

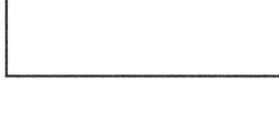

This shape is a rectangle.
It has opposite sides that are parallel and four 90° angles.

 All rectangles are parallelograms, but not all parallelograms are rectangles.

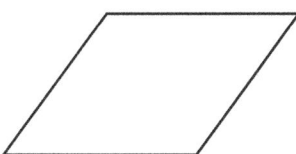

This shape is a rhombus.
It has four equal straight sides. This shape is a special parallelogram. It has opposite sides that are parallel and opposite angles that are equal.

 All squares are rhombuses, but not all rhombuses are squares.

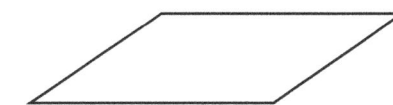

This shape is a parallelogram.
It has opposite sides that are parallel and opposite angles that are equal.

 All rhombuses are parallelograms, but not all parallelograms are rhombuses.

Classification of Quadrilaterals (continued)....

Quadrilaterals can also be classified using a Venn diagram to show relationships.

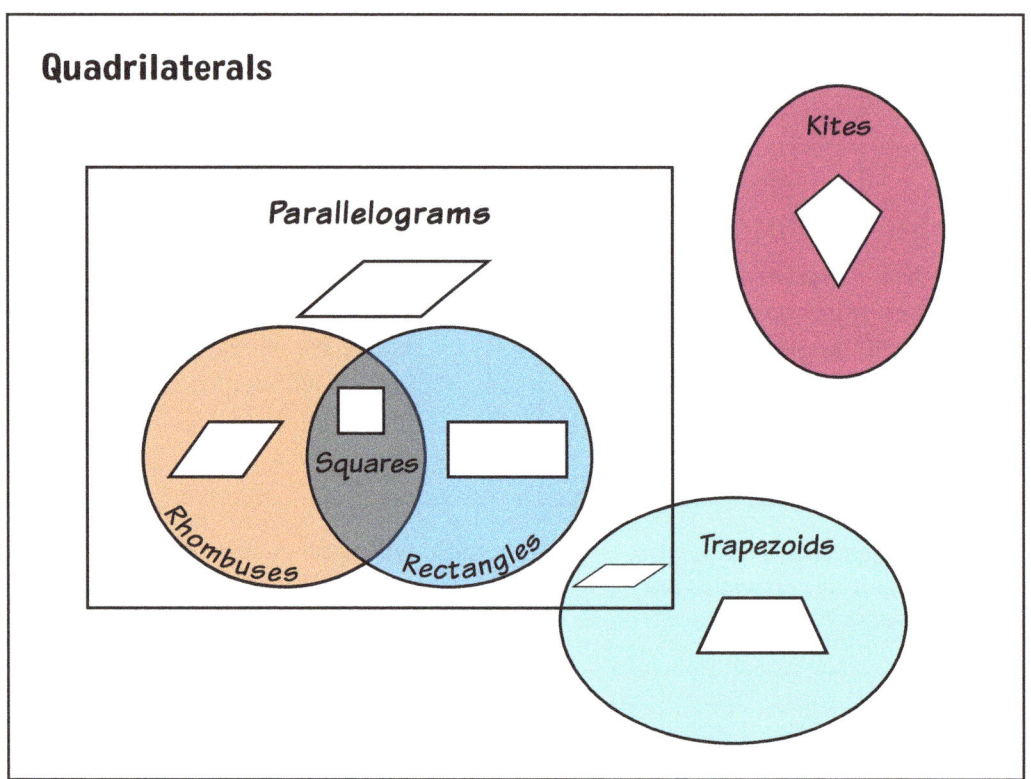

 Trapezoids have two valid definitions that differ slightly. Some define a trapezoid as a four-sided shape that has **only one** pair of parallel sides ⬜, which means that it would not be a parallelogram. Others define a trapezoid as having **at least one** pair of parallel sides ⬜, which means that it could be a parallelogram.

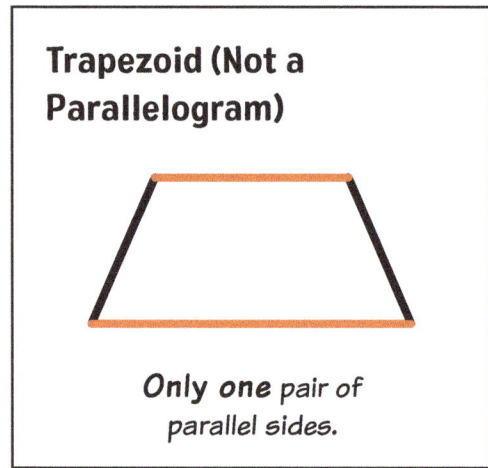

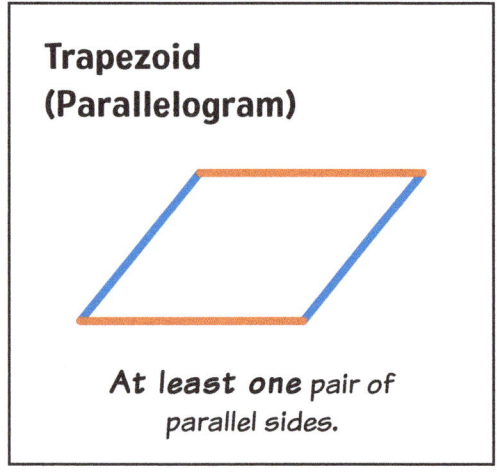

3-D Shapes

Three-dimensional shapes have **length**, **width**, and **depth**. They are not flat, but instead have volume.

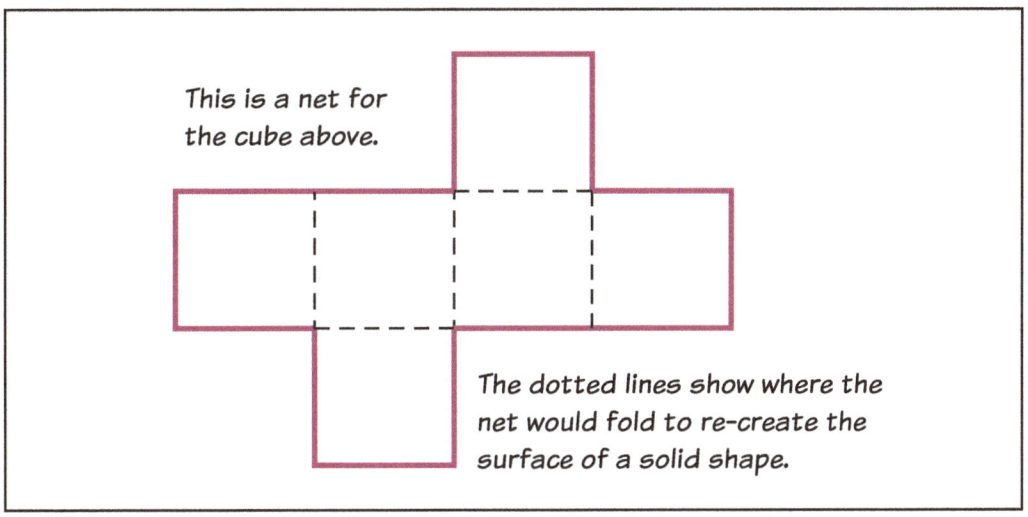

- The **face** of a solid figure (prism) is a flat surface. A prism has two faces called **bases**. Bases are parallel and the same shape and size (congruent). Usually bases are the top and bottom of the prism.

- An **edge** is a line where two faces meet.

- A **vertex** is a point where two or more lines, corners, or edges meet. **Vertices** is plural.

- A **net** is when you cut the surface of a solid shape along one or more edges and unfold it into a flat shape.

This is a net for the cube above.

The dotted lines show where the net would fold to re-create the surface of a solid shape.

3-D Shapes (continued)

Solid		Faces	Edges	Vertices	Nets (one of many)
Cube		6	12	8	
Right Rectangular Prism		6	12	8	
Right Circular Cone		2	1	1	
Right Circular Cylinder		3	2	0	
Right Rectangular Pyramid		5	8	5	

 This chart shows just one of many different ways to make a net for each of the solids.

Composite Shapes...........................

A composite shape is a shape composed from two or more other shapes. A composite shape can be either two-dimensional or three-dimensional.

2-D Composite Shapes

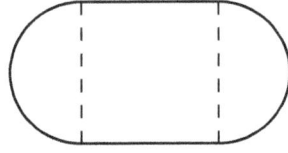

This shape was made by combining a square with two half circles.

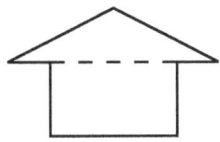

This shape was made by combining a triangle with a rectangle.

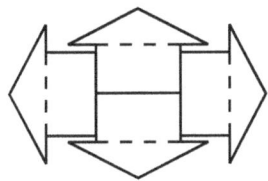

This shape was made by combining four of the arrow-shaped figures, above.

3-D Composite Shapes

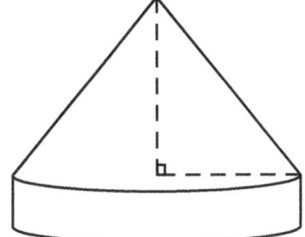

This shape was made by combining a right circular cylinder with a right circular cone.

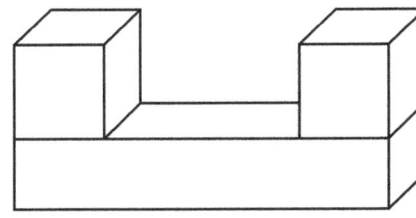

This shape was made by combining two cubes with a right rectangular prism.

Chapter 15

Measurement

Measurement allows us to quantify and compare the attributes of objects in the real world using agreed-upon standardized units. These units describe a variety of attributes including length, weight, capacity, volume, money values, units of time, temperature, and the like.

Measurement provides a way to describe "how much" of something there is in a meaningful way that others can understand. This enables clear communication and accurate calculations in various situations, from everyday life to scientific research.

Because of the incredible usefulness of measurement principles, instruction in measurement begins in the primary grades as children begin by measuring length and then move into measuring weight, volume, and capacity. It continues through the middle grades, where they learn to measure more abstract concepts like area and volume.

One aspect of measurement that presents unique challenges is the fact that in the United States and other parts of the world, there are two primary systems of measurement: U.S. customary (sometimes called "imperial") and metric. Both of these systems provide units for measuring length, weight/mass, capacity, volume, and temperature. Interestingly, the imperial system is the primary system in only three countries: the United States, Liberia, and Myanmar. The metric system is the primary system used in most other parts of the world, although some countries, such as Canada, use a mixture of the two systems.

Typical Trajectory in Most State Standards Frameworks:

- Grades K–1: Compare lengths of objects; measure length using nonstandard units; tell time to the nearest half-hour
- Grade 2: Measure, estimate, and compare lengths using U.S. customary and metric units; tell time to the nearest five minutes; value-count U.S. coins and bills
- Grade 3: Measure, estimate, and compare liquid volume and mass; tell time to the nearest minute; solve word problems involving U.S. coins and bills; calculate area and perimeter
- Grade 4: Compare units and solve problems using units within each measurement system (length, weight, liquid volume, time, money); calculate area and perimeter using formulas
- Grades 5–6: Convert among different-sized standard measurement units within a given measurement system; relate volume to multiplication and addition

Measurement: Length

The **length** of an object is measured by laying multiple copies of a shorter object end to end, with no gaps and no overlaps. Measurements can be exact or they can be estimates that are "close enough."

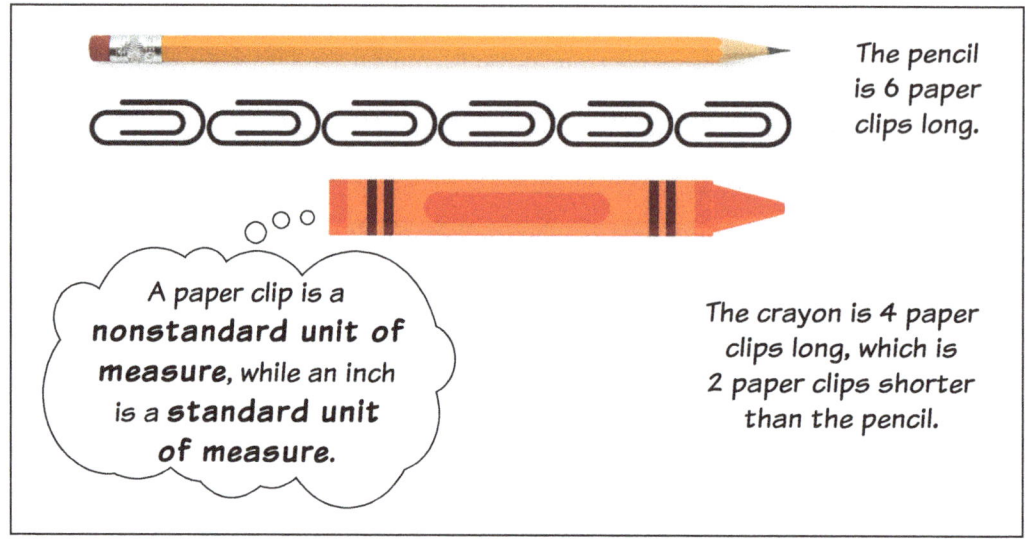

The pencil is 6 paper clips long.

A paper clip is a **nonstandard unit of measure**, while an inch is a **standard unit of measure**.

The crayon is 4 paper clips long, which is 2 paper clips shorter than the pencil.

The spoon is the shortest utensil. The knife is longer than both the fork and the spoon.

The apple is about $4\frac{1}{2}$ inches tall.

The apple is about 11 centimeters tall.

Source: Pencil image from iStock.com/Ivantsov; Paper clip images from iStock.com/isiddheshm; Crayon image from iStock.com/Artster Design; Utensils images from iStock.com/martinspurny; Apple images from iStock.com/warmworld; Ruler images from iStock/Aji Nugroho

Length Measurement: Word Problems..........

Jane wondered how much taller her framed family picture was than her favorite book.

The book is 11 inches tall

The picture frame is 16 inches tall

$$16 - 11 = 5$$

Jane's picture frame is 5 inches taller than her book.

Source: Book image from Istock.com/filo; Picture frame image from iStock.com/enviromantic; Ruler images from iStock/Aji Nugroho

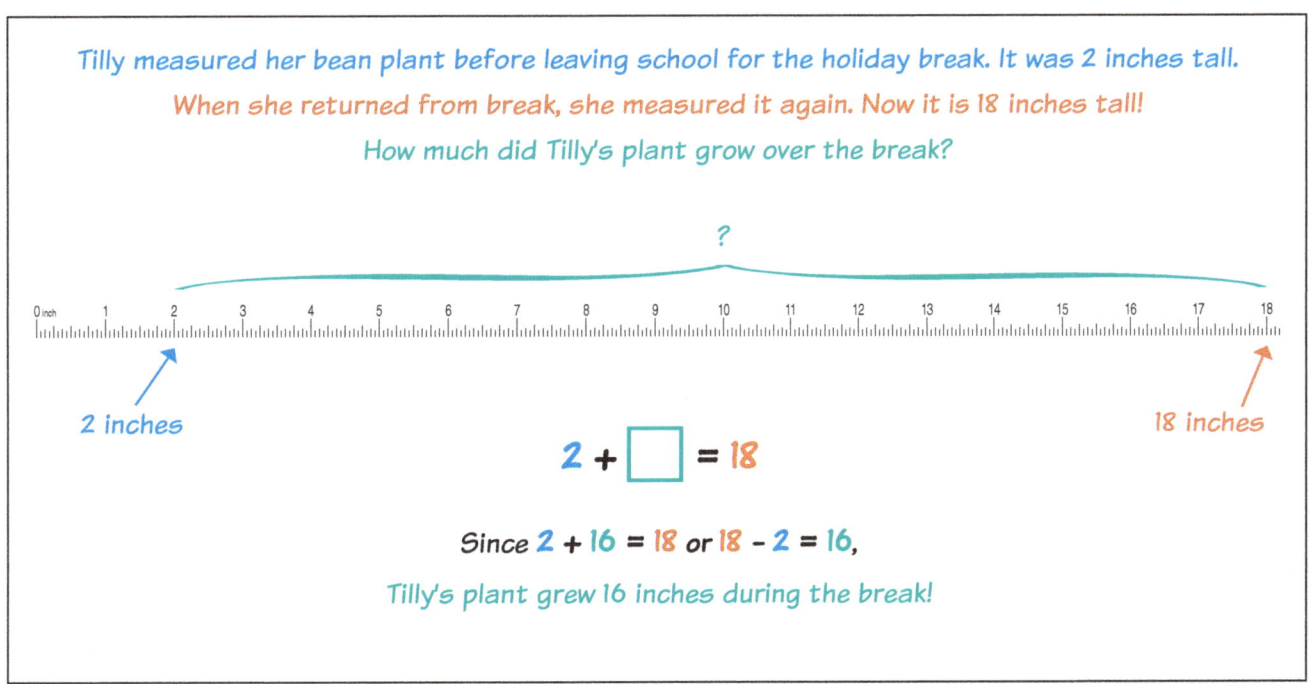

Tilly measured her bean plant before leaving school for the holiday break. It was 2 inches tall. When she returned from break, she measured it again. Now it is 18 inches tall! How much did Tilly's plant grow over the break?

2 inches

18 inches

$$2 + \square = 18$$

Since $2 + 16 = 18$ or $18 - 2 = 16$,
Tilly's plant grew 16 inches during the break!

Measurement

Length Measurement:
Word Problems (continued)

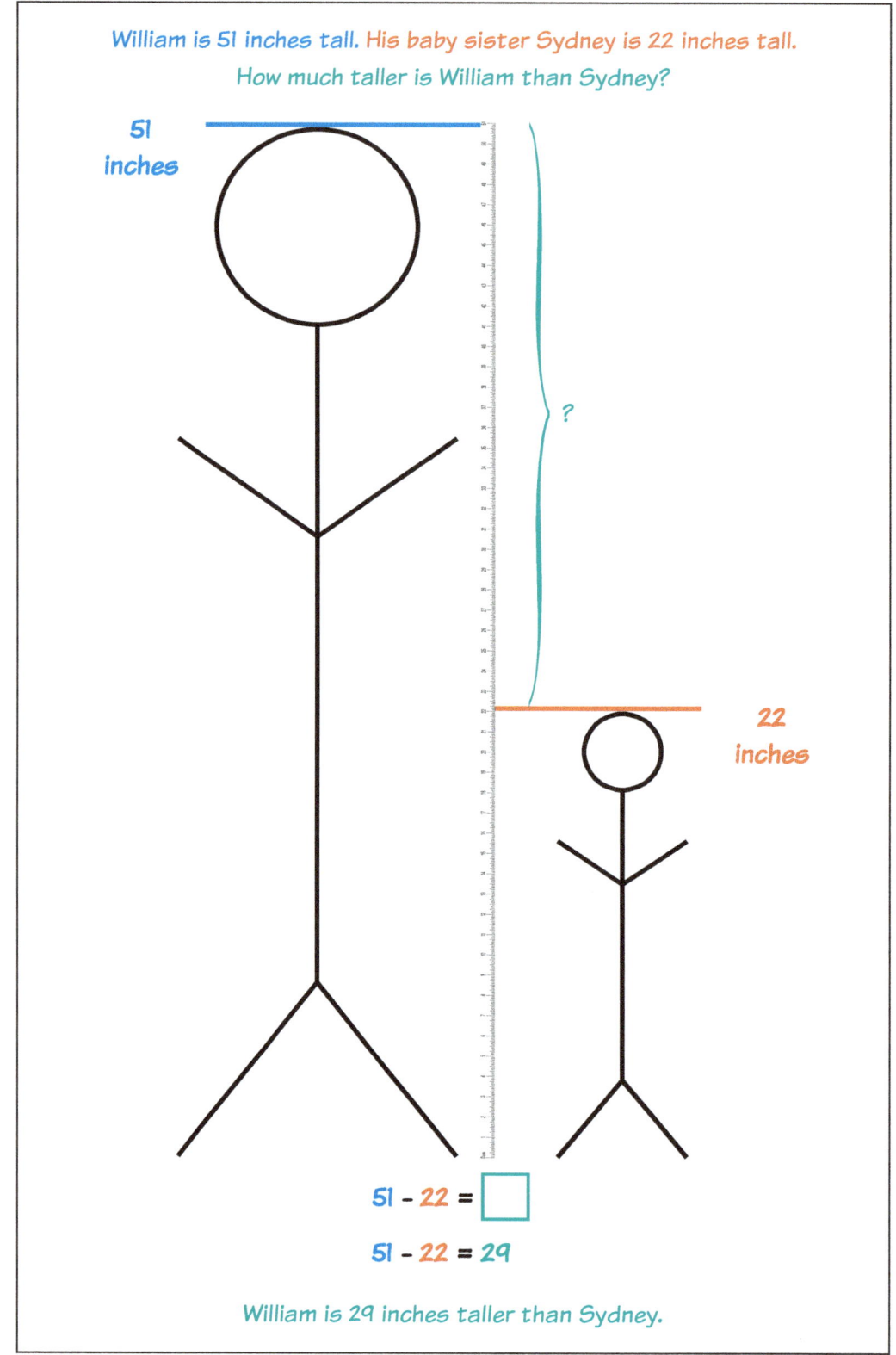

Source: Ruler image from iStock/Aji Nugroho

Length Measurement: U.S. Standard

The table below identifies how **length** is measured in the U.S. Standard/Customary System.

Unit	Equivalent	Benchmark
inch (in.)		width of a quarter
foot (ft)	12 inches	length of 3-ring notebook
yard (yd)	3 feet	from your nose to the tip of your fingers with arm outstretched
mile (mi)	5,280 feet	20-minute walking distance

Length

Fiona is cutting strips of ribbon 18 inches long. How many feet long is each strip of ribbon?

Since 1 foot = 12 inches

$$18 \div 12 = 1\frac{1}{2} \text{ feet}$$

or

$$\frac{12 \text{ in.}}{1 \text{ ft}} = \frac{18 \text{ in.}}{x}$$

 Length can be measured using a ruler, yard stick, or measuring tape.

Weight Measurement: U.S. Standard

The following table identifies how **weight** is measured in the U.S. Standard/Customary System.

Unit	Equivalent	Benchmark
ounce (oz)		slice of bread
pound (lb)	16 ounces	2 large apples
ton (T)	2,000 pounds	small car

Weight

John's new puppy weighs 4 pounds. How many ounces does his puppy weigh?

Since 1 pound = 16 ounces

$$4 \times 16 = 64 \text{ oz}$$

or

$$\frac{1 \text{ lb}}{16 \text{ oz}} = \frac{4 \text{ lb}}{x}$$

 Weight can be measured using a scale.

Liquid Volume Measurement: U.S. Standard . . . . .

The following table identifies how **liquid volume** is measured in the U.S. Standard/Customary System.

	Unit	Equivalent	Benchmark
Liquid Volume	fluid ounce (fl oz)	2 tablespoons	a serving of peanut butter
	cup (c)	8 fluid ounces	school milk container
	pint (pt)	2 cups	individual water bottle
	quart (qt)	2 pints	carton of juice
	gallon (gal)	4 quarts	jug of milk

Liquid Volume

Molly is making hot chocolate using a recipe that calls for 1 pint of milk. How many batches can she make using 1 gallon of milk?

Molly can make 8 batches of hot chocolate.

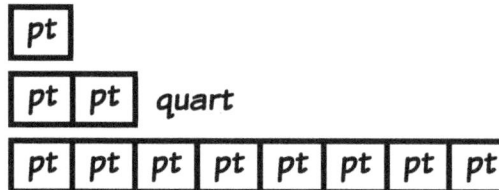

Length, Weight, and Liquid Volume Measurement: Metric

The basic metric units are **meters** (for length), **grams** (for weight) and **liters** (for liquid volume). Metric **units** are friendly to work with since they are all based on multiples of ten.

largest ←——————————————————————————→ smallest

kilo-	hecto-	deca-	unit	deci-	centi-	milli-
1,000×	100×	10×	1×	$\frac{1}{10}×$	$\frac{1}{100}×$	$\frac{1}{1000}×$

> A decagram is 10 times smaller than a hectogram and 100 times smaller than a kilogram.

> Although still smaller than a meter, a centimeter is 10 times larger than a millimeter.

 To convert from a smaller unit of measure to a larger unit of measure, **divide** by the relevant multiple of ten.

4,000 milliliters = _____ liters	675 centimeters = _____ meters
$\frac{4000}{1000} = 4\ L$	$\frac{675}{100} = 6.75\ m$

 To convert from a larger unit of measure to a smaller unit of measure, **multiply** by the relevant multiple of ten.

8 centimeters = ____ millimeters	38.2 kilometers = ____ meters
8 × 10 = 80 mm	38.2 × 1,000 = 38,200 m

Telling Time

Time is measured in **hours**, **minutes**, and **seconds** on analog and digital clocks.

Video 18: Telling Time With Fraction Pieces and Clocks: You can use fraction circle pieces and small clock faces to help students "see" half-hours and quarter-hours on a clock (thirds, sixths, and twelfths may also be included).

Pictured here: foam fraction circles and small geared clock

https://qrs.ly/rqg99ma

Analog Clocks

Analog clocks have two or three hands. The hour hand is the shortest, the minute hand is longer, and the second hand is long and thin. These hands all move **clockwise**, or to the right, when starting at the 12.

Exact Hour	Between Hours	Exact Minute
Shown when the minute hand is pointing directly at 12 and the hour hand is pointing directly at any number on the clock	Unless the minute hand is on the 12, the hour is always the number counter-clockwise (left starting at 12) from the hour hand.	There are 5 minutes between each number on the clock face, so each hour mark represents 5 times as many minutes.

Eleven o'clock   Nine twenty   Six oh nine

Minutes can also be read as fractions of an hour.

Quarter past 7 o'clock   **Half past** 10 o'clock   **Quarter till** 8 o'clock

Source: Clock images from iStock.com/FARBAI

Digital Clocks

On a digital clock, the hour is always the number to the left of the colon.

 Eleven thirty-two at night

- From midnight until before noon, the time is a.m.
- From noon until before midnight, the time is p.m.

Elapsed Time

The time that passes between a start time and an end time is called **elapsed time**. To find elapsed time, count from the starting time to the finishing time.

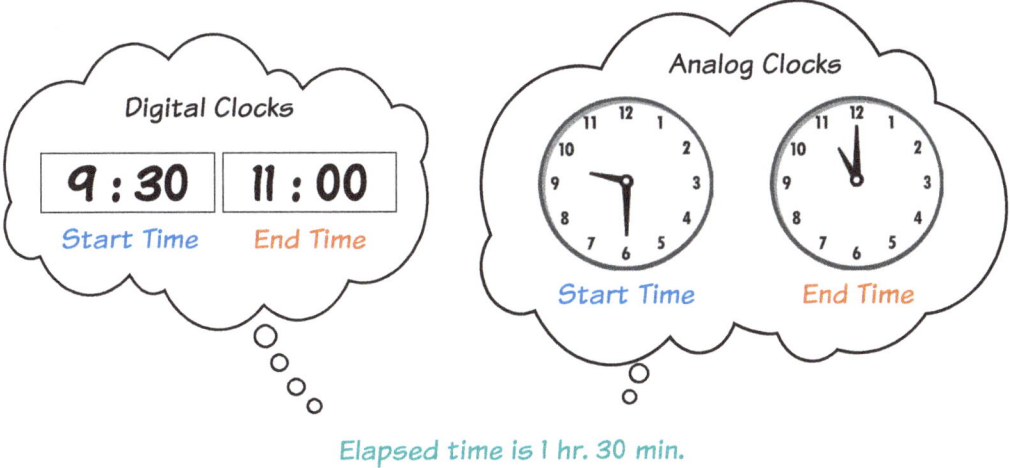

Elapsed time is 1 hr. 30 min.

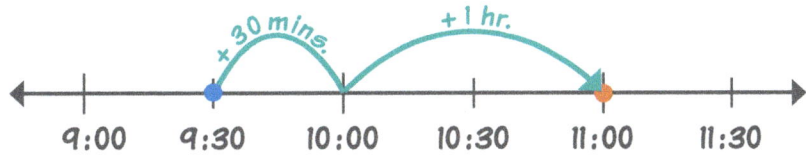

Will went to a movie that started at 11:30 a.m. and lasted 2 hours, 15 minutes. What time did the movie finish?

	Elapsed Time
11:30	(movie starts)
11:30–12:00	$\frac{1}{2}$ hour
12:00–1:00	1 hour
1:00–1:30	$\frac{1}{2}$ hour
1:30–1:45	15 minutes

} 2 hours

The clock will begin its count over again when it reaches noon!

The movie ended at 1:45 p.m.

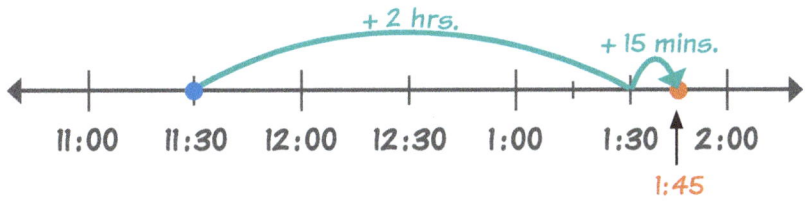

Source: Clock images from iStock.com/FARBAI

Measurement

Money

Video 19: Proportional Coin Cards: You can use proportional coin cards to help students "see" comparative coin values and to value-count mixed coins. *Pictured here: KP® Dollar Board and coin cards—click on link below to download.*

https://qrs.ly/8jg99md

Link for *KP® Dollar Board and coin cards* download:

https://qrs.ly/x5g99ju

The U.S. Monetary System is conveniently based on groups of ten, allowing us to write monetary amounts using ***decimal notation***.

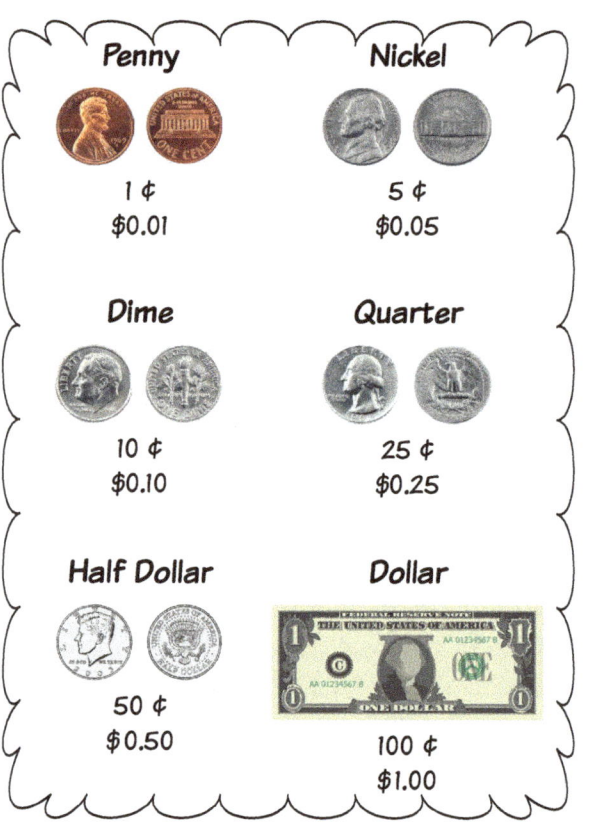

If you have 2 dimes and 3 pennies, how many cents do you have?

1¢ + 1¢ + 1¢ = 3¢

10¢ + 10¢ = 20¢

3¢ + 20¢ = 23¢ or $0.23

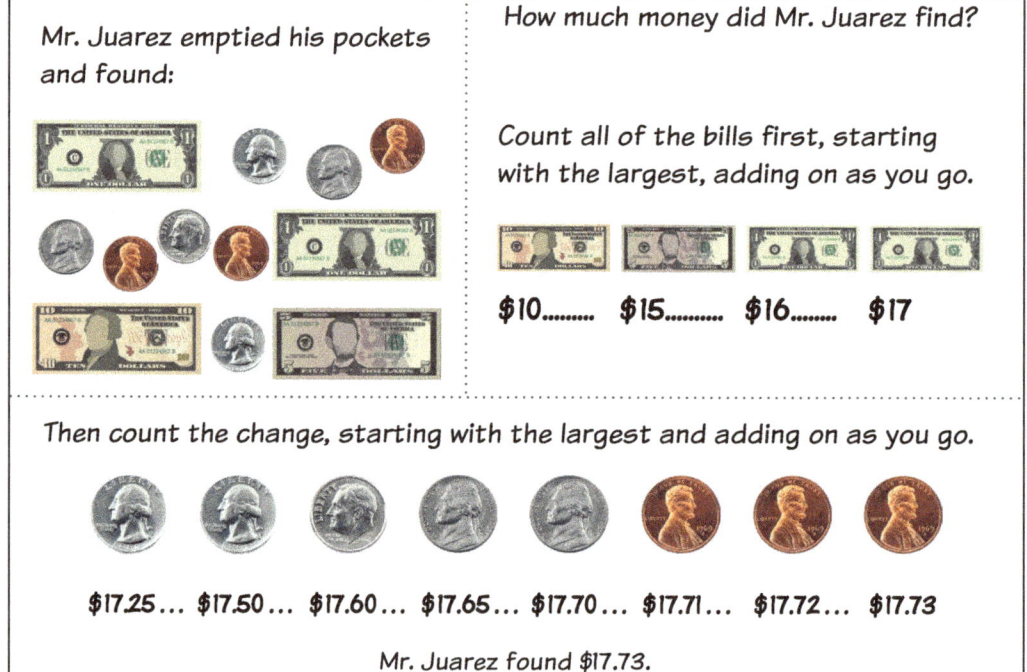

Source: Bill images by istock.com/SereiKorolko; Coin images by istock.com/filo

Perimeter and Area

To find the **perimeter** of any shape, you simply add the lengths of all sides (even the sides that do not show the measurement). There are different formulas to find the **area** for different shapes.

> **Formulas for Rectangles**
> Perimeter = 2 (l + w)   l = length
> Area = l × w   w = width

Perimeter

Perimeter is the distance around a polygon.

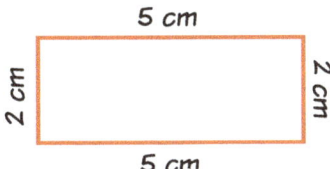

P = 5 cm + 2 cm + 5 cm + 2 cm
P = 14 cm

Area

Area of a figure is the number of square units inside the figure without gaps or overlaps.

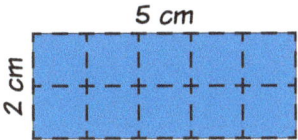

A = 2 cm × 5 cm
A = 10 cm² (square centimeters)

When fencing a garden, you need to calculate **perimeter**.

P = 11 ft + 24 ft + 11 ft + 24 ft
P = 70 ft

When covering a floor with carpet or tile, you need to calculate **area**.

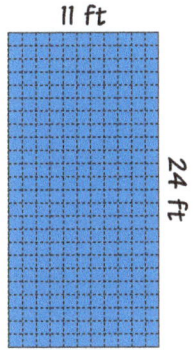

A = 11 ft × 24 ft
P = 264 ft² (square feet)

Perimeter and Area of Rectangles With Fractional Side Lengths

To find the **perimeter** and **area** of rectangles without whole number side lengths, you can follow the same procedure as you would for whole numbers.

Perimeter

Perimeter is the distance around a polygon.

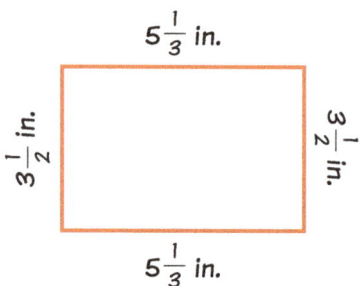

$P = 2(l) + 2(w)$

$P = 2(3\frac{1}{2}) + 2(5\frac{1}{3})$

$P = 7 + 10\frac{2}{3}$

$P = 17\frac{2}{3}$ inches

Area

Area of a figure is the number of square units inside the figure without gaps or overlaps.

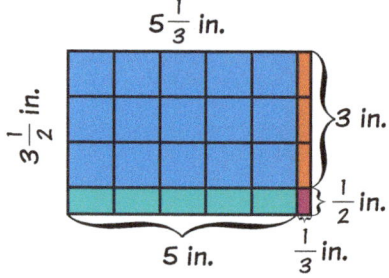

$A = l \times w$

$A = 5\frac{1}{3} \times 3\frac{1}{2}$

I'll use the decomposition strategy from page 91!

$(5 + \frac{1}{3}) \times (3 + \frac{1}{2})$

$A = (5 \times 3) + (5 \times \frac{1}{2}) + (3 \times \frac{1}{3}) + (\frac{1}{2} \times \frac{1}{3})$

$A = 15 + 2\frac{1}{2} + 1 + \frac{1}{6}$

$A = 18\frac{4}{6}$ in.² (square inches)

Area of a Parallelogram

If you cut up a parallelogram, you can rearrange the pieces to make a rectangle. This way you can use what you know about finding the **area** of a rectangle to find the **area** of a parallelogram.

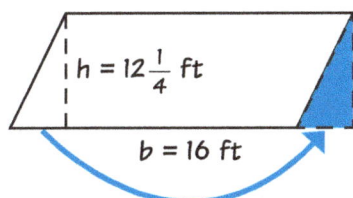

Since the area of a rectangle is length times width (or base times height), the area of this parallelogram is also base times height.

$$A = b \times h$$
$$= 16 \times 12\frac{1}{4}$$
$$= 196 \text{ ft}^2$$

b = base
h = height

Area of a Triangle

Every triangle is half a rectangle. Since the **area** of a rectangle is length times width (or base times height), the **area** of a triangle is half that product.

h = 5 in.
b = 4 in.

b = 10 cm
$h = 9\frac{1}{2}$ cm

$$A = \frac{1}{2} \times b \times h$$

$\frac{1}{2} \times 4 \times 5$
$\frac{1}{2} \times 20$
$= 10 \text{ in.}^2$

$\frac{1}{2} \times 10 \times 9\frac{1}{2}$
$\frac{1}{2} \times 95$
$= 47\frac{1}{2} \text{ cm}^2$

Solving Area Problems

If you know that the **area** of a parallelogram is bh (base times height or length times width) and that the **area** of a triangle is $\frac{1}{2}bh$ ($\frac{1}{2}$ of the base of the triangle times the height), you can use this information to solve other area problems.

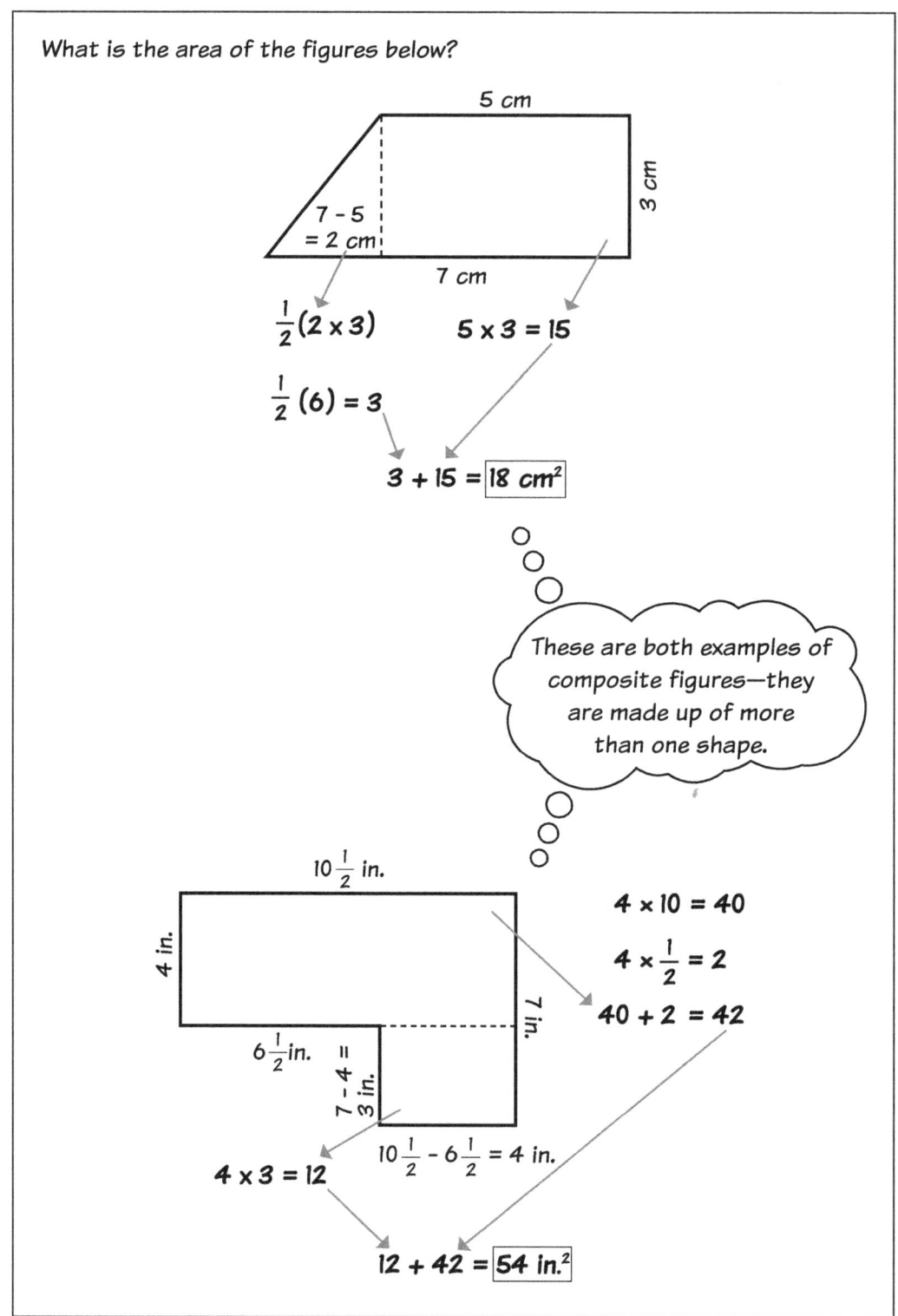

Surface Area

Surface area (SA) is the total area of all of the faces of an object. We measure it in square units. You can think of surface area as the amount of wrapping paper that would be needed to completely wrap an object without any gaps or overlaps.

You can find the surface area of any object by finding the area of each face and adding them together.

Video 20: Surface Area With Linking Cubes and Nets: You can use linking cubes and nets to help students "see" surface area. *Pictured here: multi-link cubes and paper nets*

https://qrs.ly/bpg99mh

What is the surface area of the objects below?

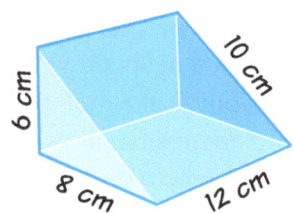

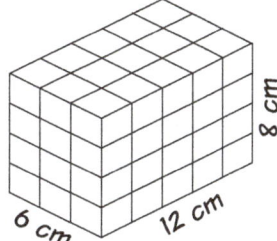

There are two triangle faces and three rectangular faces on the triangular prism.

Area of a triangle is $\frac{1}{2}(l \times w)$.

$\frac{1}{2}(6 \times 8) + \frac{1}{2}(6 \times 8) = 48$ cm²

The area of each rectangle is found by multiplying $l \times w$.

$(12 \times 8) + (12 \times 6) + (10 \times 12)$
$= 96 + 72 + 120$
$= 288$ cm²

The total of all faces is

$48 + 288 = 336$

SA = 336 cm²

There are three different rectangular faces. Each has a matching face on the opposite side of the rectangle for a total of six faces in all. Area of a rectangle is $l \times w$.

$2(8 \times 12) + 2(8 \times 6) + 2(6 \times 12)$

$2(96) + 2(48) + 2(72) =$
$192 + 96 + 144 = 432$

SA = 432 cm²

Volume..

The **volume** is the amount of space inside a three-dimensional (**3-D**) shape. Volume is measured in cubic units (u^3), which tells you how many cubes would fit inside the prism, like blocks in a box without gaps or overlaps.

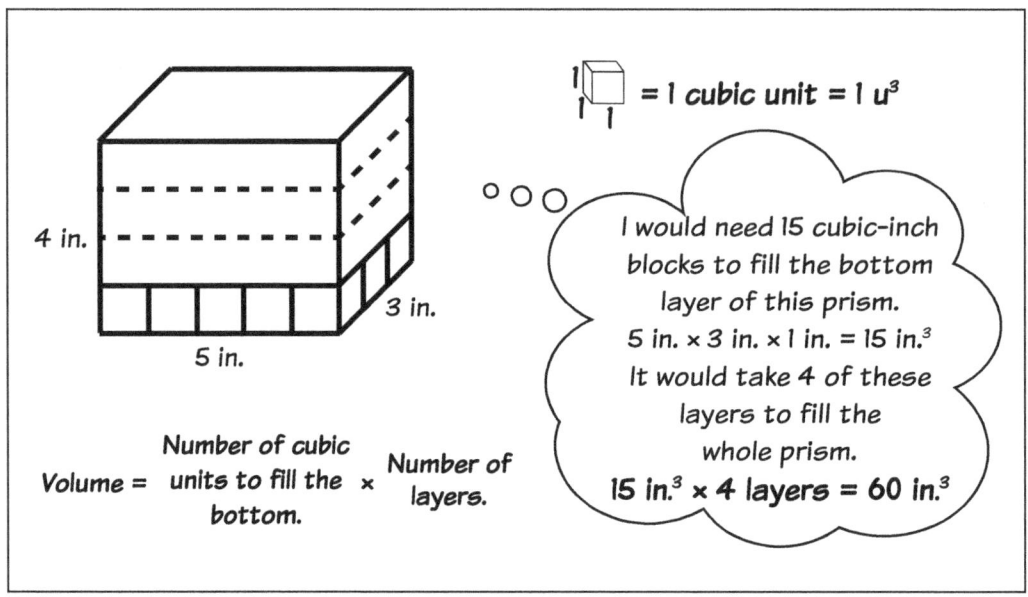

Volume = Number of cubic units to fill the bottom. × Number of layers.

I would need 15 cubic-inch blocks to fill the bottom layer of this prism.
5 in. × 3 in. × 1 in. = 15 in.³
It would take 4 of these layers to fill the whole prism.
15 in.³ × 4 layers = 60 in.³

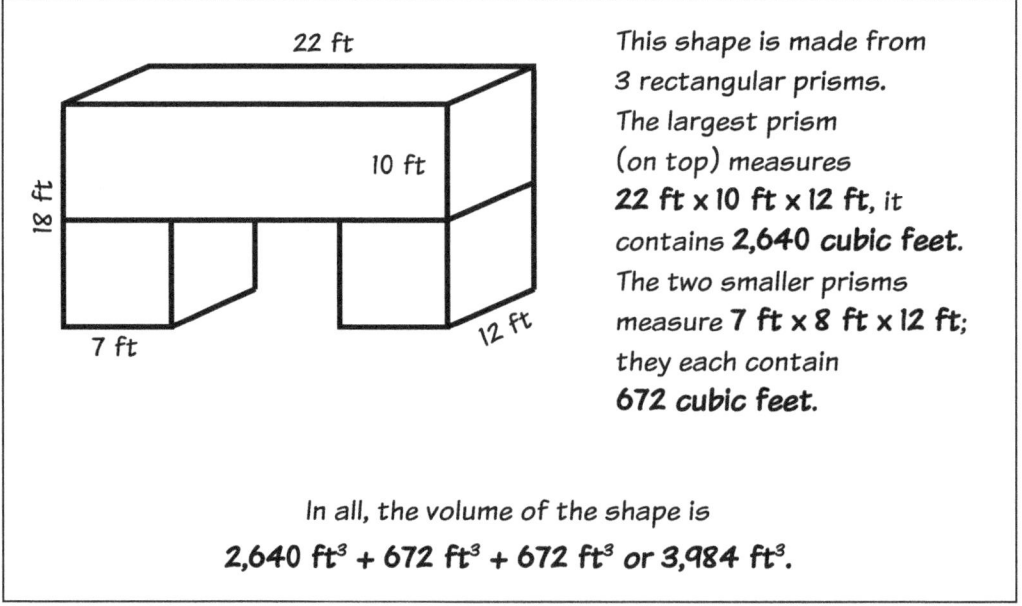

This shape is made from 3 rectangular prisms. The largest prism (on top) measures **22 ft × 10 ft × 12 ft**, it contains **2,640 cubic feet**. The two smaller prisms measure **7 ft × 8 ft × 12 ft**; they each contain **672 cubic feet**.

In all, the volume of the shape is
2,640 ft³ + 672 ft³ + 672 ft³ or 3,984 ft³.

Volume (continued)

$$V = l \times w \times h$$
length × width × height

OR

$$V = B \times h$$
Area of base × height

Video 21: Volume With Linking Cubes and Graph Paper: You can use linking cubes and graph paper to help students "see" different interpretations for volume. *Pictured here: multi-link cubes for the "copies of layers" interpretation and graph paper for the "area of the base" interpretation*

https://qrs.ly/89g99mj

Two different boxes of cereal cost the same amount. Wanting the most for her money, Sofia wants to know which box holds more cereal.

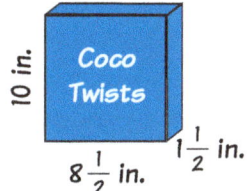

$8\frac{1}{2}$ in. × $1\frac{1}{2}$ in. × 10 in.

$(1 \times 10) + (\frac{1}{2} \times 10)$
$10 + 5$
15

$= 8\frac{1}{2} \times 15$

$(8 \times 15) + (\frac{1}{2} \times 15)$
$120 + 7\frac{1}{2}$
$= 127\frac{1}{2}$ in.³

7 in. × $2\frac{1}{2}$ in. × 10 in.

$(7 \times 2) + (7 \times \frac{1}{2})$
$14 + 3.5$
17.5

$= 17.5 \times 10$
$= 175$ in.³

The Lucky Leaves box contains more cereal.

Chapter 16

Data

In the real world, understanding data has always been important for making informed decisions, but recently the significance of understanding data has increased exponentially. Now, due to the increase of data used in daily life, in the workplace, in government, as well as in artificial intelligence integration, students' development of data literacy has become an essential life skill.

The work children do with data in primary and middle grades develops their data literacy. As students develop data literacy, they come to understand that the same set of data can tell multiple stories, and it's in their best interest to become flexible in "reading" these data stories.

Because everyone is both a creator and user of data in our modern world, student experiences in data should be explored through rich and meaningful contexts that are relevant to the students and their communities. Students should have multiple opportunities throughout the school year to design data experiments, compose questions, collect data, and finally interpret and analyze the data set to see what stories it tells. These can be meaningful and rewarding experiences for children.

Mathematically, data experiences will draw upon numerical, computational, and statistical concepts that prepare students to be thoughtful consumers and producers of data. Furthermore, working with data and statistics affords students the opportunity to apply the many skills they develop in the other math domains, such as algebraic thinking, number and operations in base ten, number and operations with fractions, measurement, and geometry.

Typical Trajectory in Most State Standards Frameworks:

- Grade K: Count and classify mixed sets of objects
- Grades 1–2: Organize, represent, and interpret data, including measurement data; create picture graphs and bar graphs to represent data
- Grades 3–5: Create scaled picture and bar graphs to represent data; make line plots to display data sets, including fractional data sets
- Grades 6+: Examine variability, distribution, measures of center (mean, median, mode) in data sets; display and interpret data sets by creating line plots, histograms, dot plots, and box plots

Data

There are two distinct types of **data**, measurement data and categorical data.

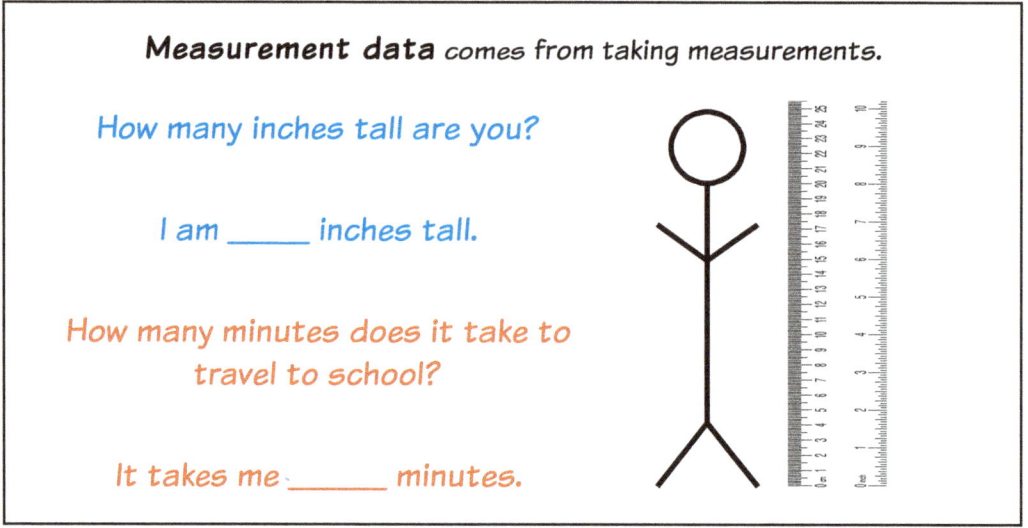

Measurement data comes from taking measurements.

How many inches tall are you?

I am _____ inches tall.

How many minutes does it take to travel to school?

It takes me _____ minutes.

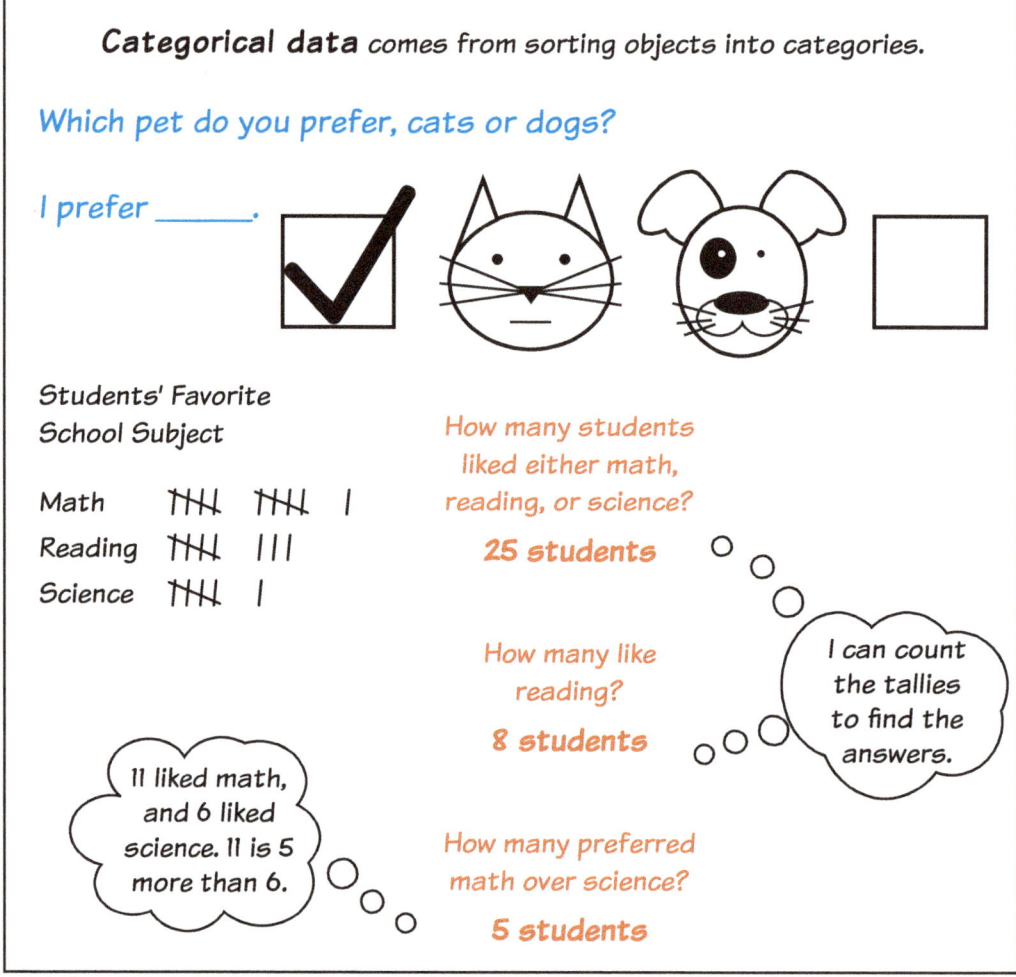

Categorical data comes from sorting objects into categories.

Which pet do you prefer, cats or dogs?

I prefer _____.

Students' Favorite School Subject

Math ||||| ||||| |
Reading ||||| |||
Science ||||| |

How many students liked either math, reading, or science?

25 students

How many like reading?

8 students

How many preferred math over science?

5 students

I can count the tallies to find the answers.

11 liked math, and 6 liked science. 11 is 5 more than 6.

Source: Cat and dog icons from iStock.com/wakashi1515; Ruler image from iStock/Aji Nugroho

Picture Graphs...........................

Picture graphs (also called pictographs) use pictures or symbols to display data in a way that is eye-catching. Pictures can be used to represent any quantity.

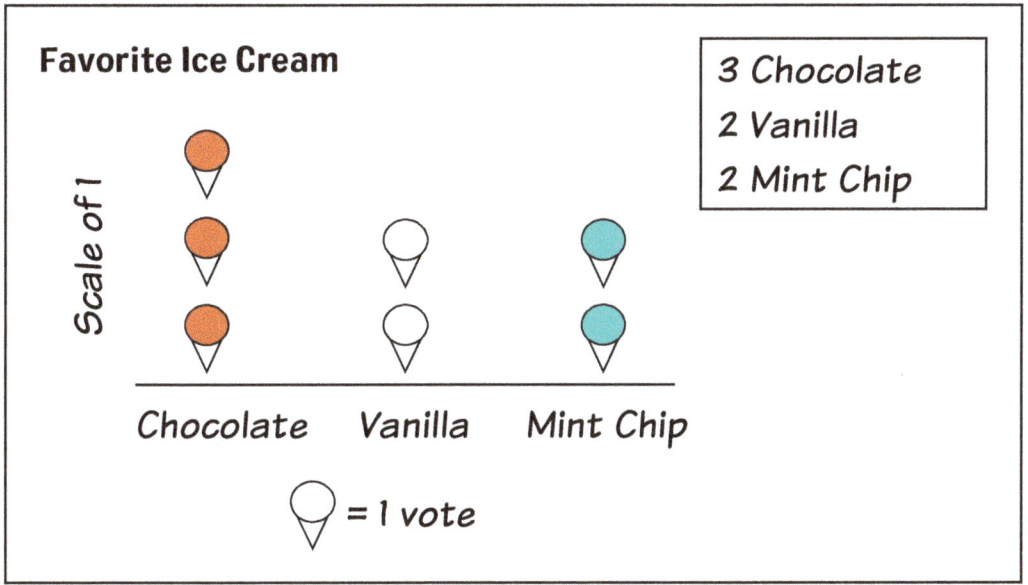

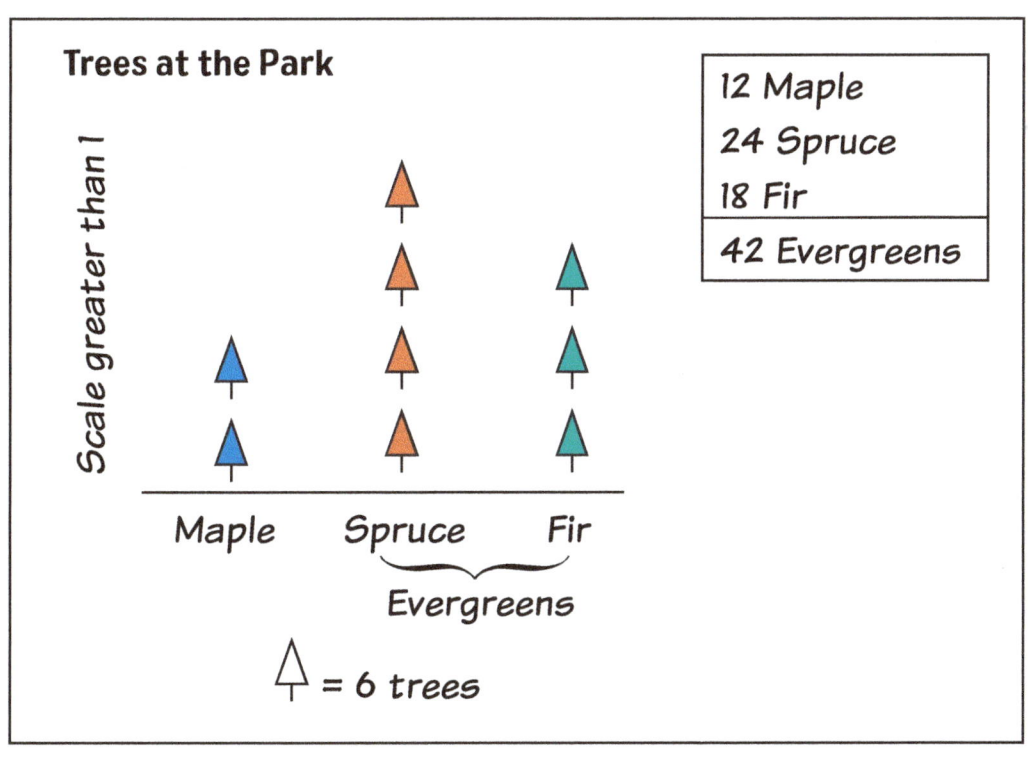

Bar Graphs

Bar graphs are used to display data that can be easily counted or measured.

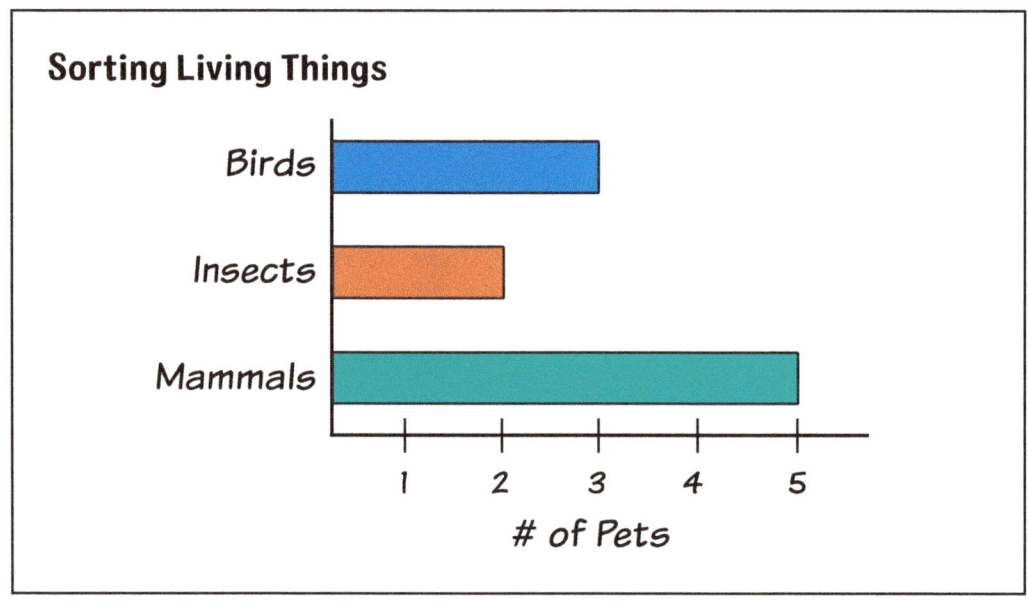

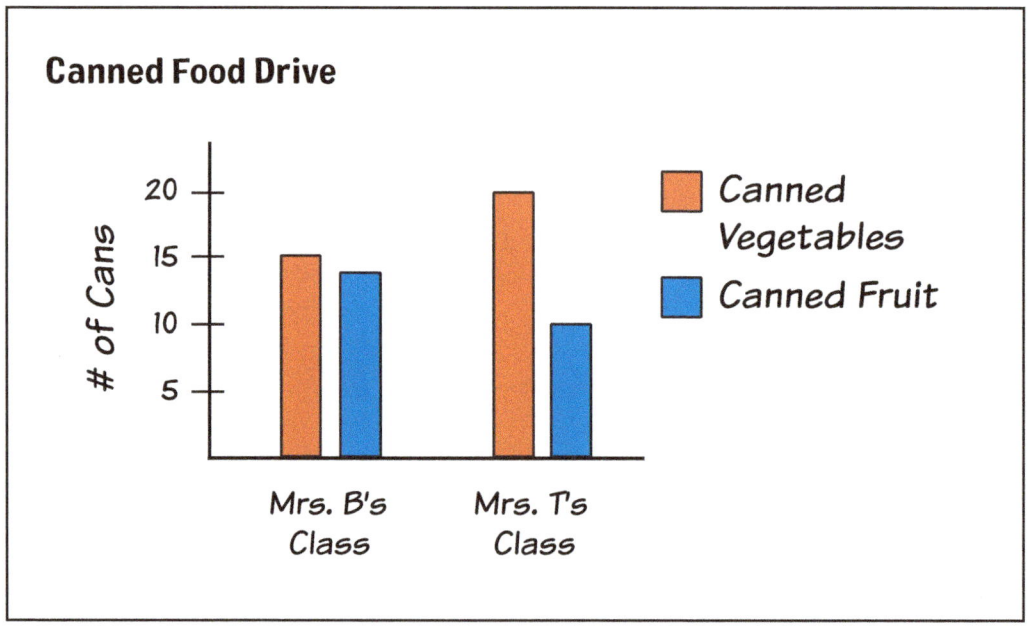

 Bar graphs do not touch because the bars represent different categories.

Line Plots (Dot Plots)

Line plots are used to help the viewer quickly identify the **range** (difference between the greatest and smallest value) and the **mode** (most frequently occurring value) of the data. Each dot represents exactly one of the items being measured.

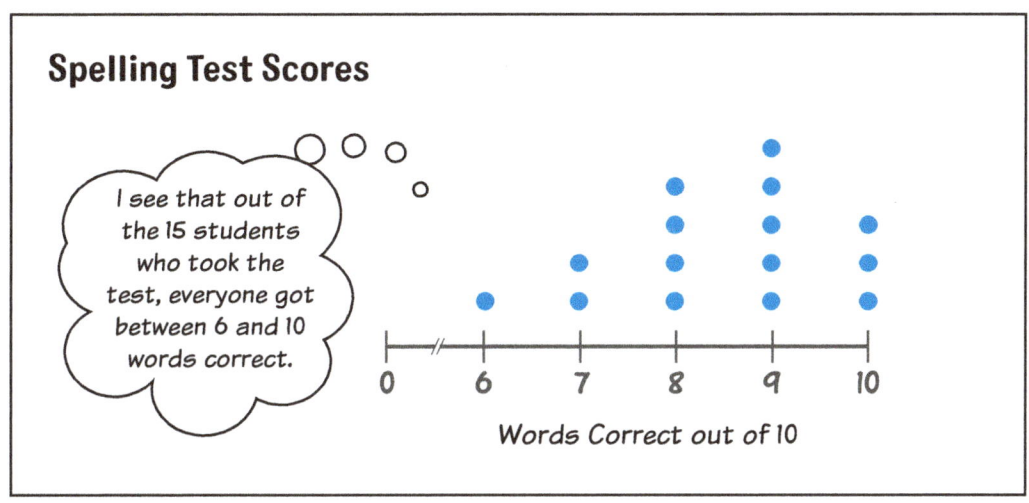

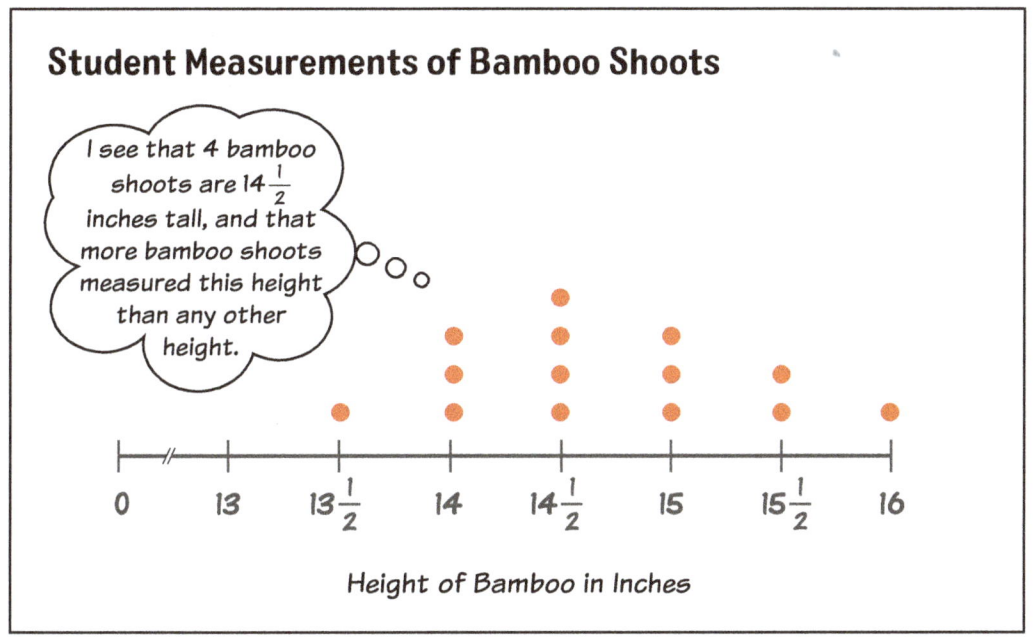

 Notice that although the number line starts at zero, the lowest data value is $13\frac{1}{2}$. You do not need to include all numbers between zero and the lowest data value, and the double slash indicates this.

Histograms

Histograms share characteristics with bar graphs. However, histograms show continuous data using touching bars, while bar graphs compare categories using separate bars.

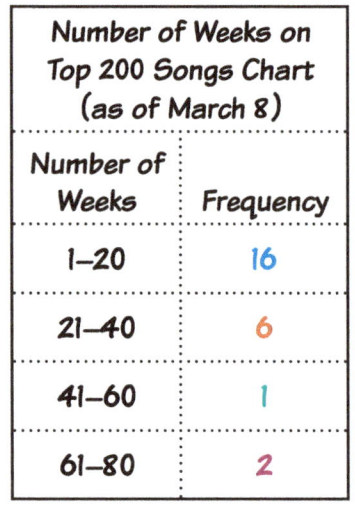

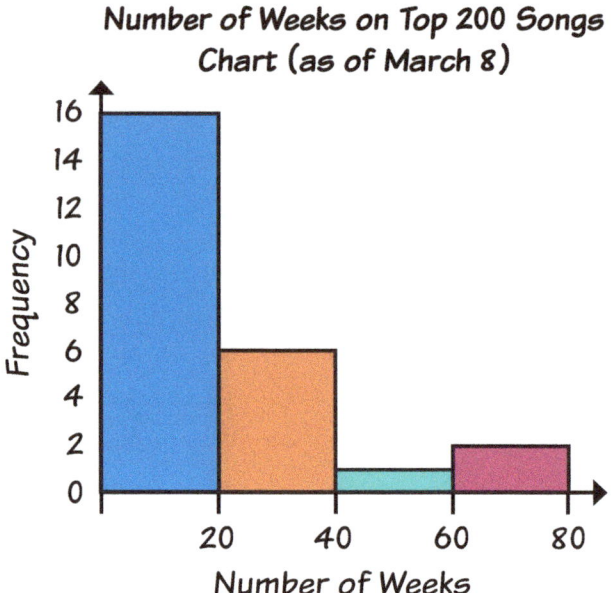

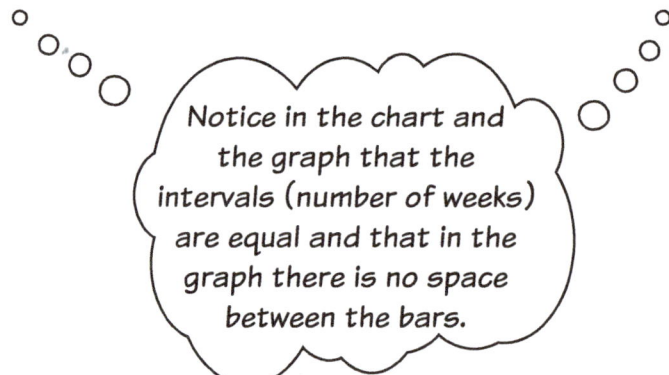

Notice in the chart and the graph that the intervals (number of weeks) are equal and that in the graph there is no space between the bars.

Systematic Listing and Counting..............

When combining different types of objects, it is sometimes necessary to find out how many different combinations can be created. The number of possible combinations can be found using organized lists or multiplication.

How many ways can you combine the articles of clothing shown below using exactly one type of shirt, one type of pants, and one type of shoes?

Shirts: Black shirt,  White shirt,  Gray shirt

Pants: Khaki pants, Jeans

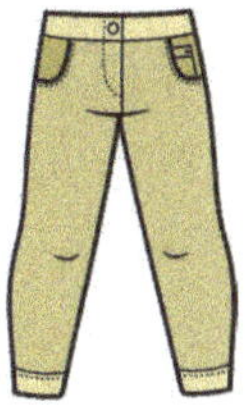

Shoes: Tennis shoes, Sandals

I can make a list, chart, or even a tree diagram and then count all of the possible combinations.

There are 12 possible outfits!

	1	2	3	4	5	6	7	8	9	10	11	12
Black shirt	X	X	X	X								
White shirt					X	X	X	X				
Gray shirt									X	X	X	X
Khaki pants	X	X			X	X			X	X		
Jeans			X	X			X	X			X	X
Tennis shoes	X		X		X		X		X		X	
Sandals		X		X		X		X		X		X

Source: black, white, and gray shirt icons by istock.com/RobinOlimb; khaki pants icon by istock.com/azul; jeans by istock.com/Photoplotnikov; tennis shoe icon by istock.com/badidok; sandals icon by istock.com/designer29

 When there are 2 or more choices to make, you can also find the number of possible combinations by simply multiplying the number of options for each choice.

3 shirts × 2 pants × 2 shoes = 3 × 2 × 2 = **12 possible outfits**

Statistical Variability

Variability refers to how "spread out" a group of measurements is. To be a **statistical question**, one would expect the answers to be "spread out" or to show variability in the answers.

For example: The question, "How old are you?" has only a single numerical answer. When asked of all the kids in a school, the question, "How old are you?" becomes a **statistical question**. The second question has many numerical answers. It has diversity or **variability** because a variety of answers are possible.

A set of data collected to answer a statistical question has a **distribution**, which can be described by its **center**, **spread**, and **overall shape**.

> The **center** can be described by a single number that summarizes all of the values in the data set.
>
> Mean, median, and mode are common measures of center.

> The **spread** or **variability** can be described by a single number that shows how the values in the data set vary.
>
> Range, interquartile range, and mean absolute deviation are all measures of variation.

> The **overall shape** can be used to describe the set of data after it is displayed graphically.
>
>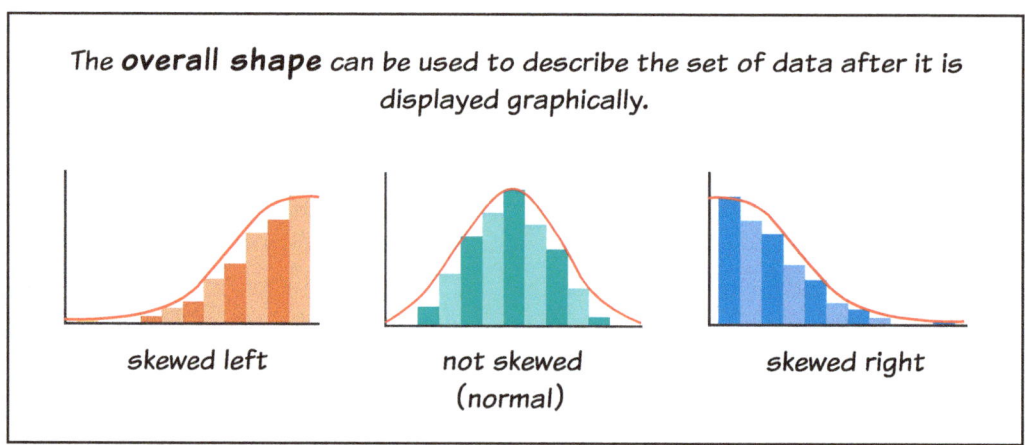
>
> skewed left   not skewed (normal)   skewed right

190   SEEING THE MATH YOU TEACH, GRADES K–6

Measures of Center/What Is Typical?

To analyze data, you can look at what is typical or **average** within the data.

> Data Set
> 17, 20, 15, 17, 16

- The **mean** is the average of your data. You find it by adding all the values together and then dividing by the number of values. The mean is affected by extreme values, so it may not be the best measure of center to use in a skewed distribution.

Mean

$$\begin{array}{r} 17 \\ 20 \\ 15 \\ 17 \\ +16 \\ \hline 85 \end{array} \qquad 5\overline{)85}^{\,17}$$

- The **median** is another measure of center. You find it by putting the data in order from least to greatest and then finding the middle number.

Median

15, 16, (17), 17, 20

- When there isn't exactly one number in the middle, you add the two middle numbers together and divide by 2. The result will be the median. The median is probably the best measure of center to use in a skewed distribution.

Median (two middle numbers)

15, 16, (17, 18), 19, 23

$$\frac{17 + 18}{2} = \frac{35}{2} = 17\frac{1}{2}$$

- The **mode** is the value that occurs most often. There can be one mode, more than one mode, or no mode at all. The mode is probably not affected by extreme values since it's unlikely the extreme values are the most common.

Mode

15, 16, (17, 17), 20

Measures of Variation/The Spread of Data.....

Measures of variation are important. They allow the reliability of measures of center to be evaluated. To describe a set of data appropriately, its variability must be described, otherwise there is the potential for data deception. Measures of variation also help guide how a graph of the data should look.

- The **range** is the difference between the biggest (greatest) and smallest (least) numbers in your data.

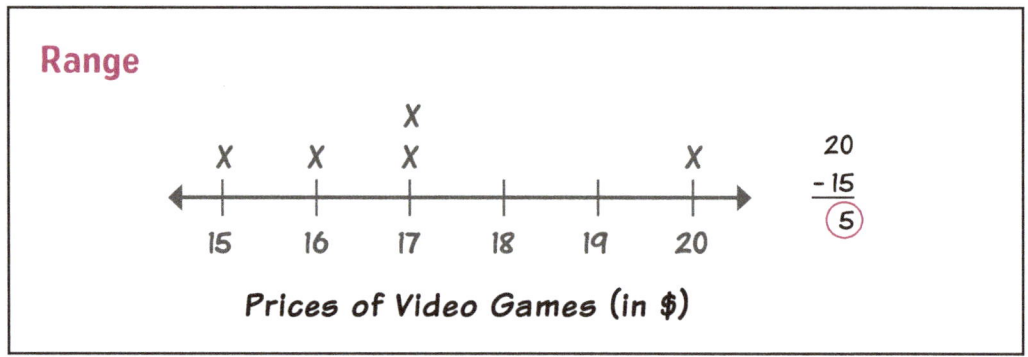

- The **interquartile range** is the range of the middle 50% of the values in a set of data. It is often used to construct a box plot to display the data (see page 194).

- The **mean absolute deviation** is the average distance between each data value and the mean of the set of data (see page 195).

Box Plots (Definitions).....................

Box plots (sometimes called **box-and-whisker plots**) are used to show how data is distributed along a number line.

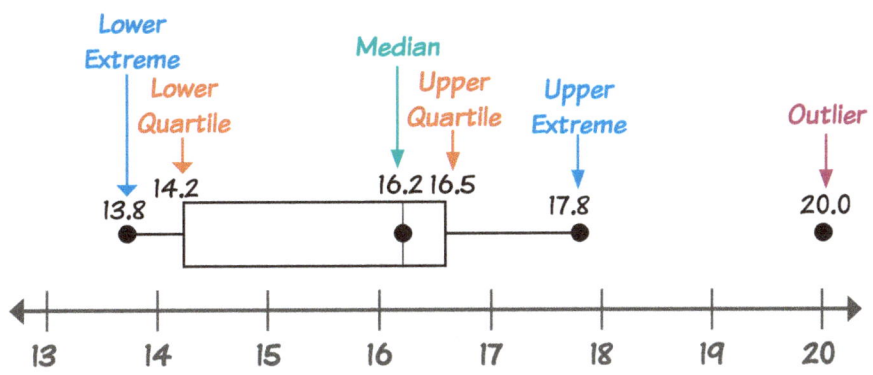

- **Lower extreme** — The lowest (smallest) piece of data.

- **Lower quartile** — The median of the lower half of an ordered set of data.

- **Median** — The middle number of a set of numbers arranged from least to greatest (see page 190 for more about median).

- **Upper quartile** — The median of the upper half of an ordered set of data.

- **Upper extreme** — The highest (largest) piece of data.

- **Outlier** — Any piece of data that is much smaller or larger than most of the other numbers in a set of data.

<u>To find outliers:</u>

① Subtract the lower quartile from the upper quartile to find the box length (16.5 − 14.2 = 2.3).

② Multiply the box length by 1.5 (2.3 × 1.5 = 3.45).

③ Add the answer from step ② to the upper quartile (16.5 + 3.45 = 19.95).

④ Subtract the answer from step ② from the lower quartile (14.2 − 3.45 = 10.75).

⑤ Outliers will be any value greater than the answer from step ③ (19.95) or less than the answer from step ④ (10.75).

Data   193

Box Plots (Example)

A **box plot** displays the **median**, **quartiles**, and **outliers** in a set of data but does not display specific values. It shows how the middle values are spread out and if any data is too far from the middle.

Creating a Box Plot

The local movie theater recorded their ticket sale earnings for seven days. The data is displayed in the table below:

Day	Earnings
1	$400
2	$450
3	$500
4	$625
5	$650
6	$580
7	$600

① Write the data in order from least to greatest:

400, 450, 500, 580, 600, 625, 650

② Draw a number line appropriate to the data with equal intervals (shown below).

③ Mark the median – 580.

④ Look at the upper half and mark the median of the upper half (upper quartile) – 625.

⑤ Look at the lower half and mark the median of the lower half (lower quartile) – 450.

⑥ Mark the upper and lower extremes (the smallest and largest numbers) – 650, 400.

⑦ Draw a box between the upper quartile and lower quartile. Split the box by drawing a line through the median. Draw two "whiskers" from the quartiles to the extremes.

Local Movie Ticket Sales

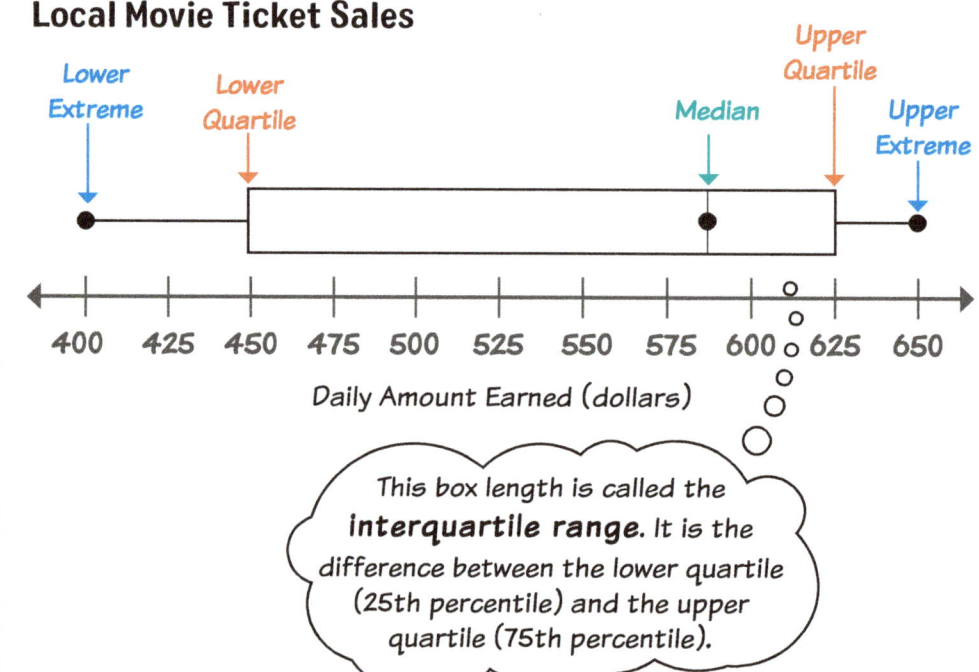

This box length is called the **interquartile range.** It is the difference between the lower quartile (25th percentile) and the upper quartile (75th percentile).

Mean Absolute Deviation

The **mean absolute deviation** is the average distance between each data value and the mean of the set of data. It is one way to find how consistent a set of data is. It is a measure of variability.

How to determine the mean absolute deviation:

① Find the mean (average): add all the numbers and divide by the number of values.

② Subtract the mean from each data point.

③ Take the absolute value of each difference (so they are all positive numbers).

④ Find the mean of the absolute values.

William's Test Scores
80, 100, 94, 88, 90, 94

① Mean of William's test scores:
$$\frac{80 + 100 + 94 + 88 + 90 + 94}{6} = 91$$

②

Data Point	Mean	Difference
80 −	91 =	−11
100 −	91 =	9
94 −	91 =	3
88 −	91 =	−3
90 −	91 =	−1
94 −	91 =	3

③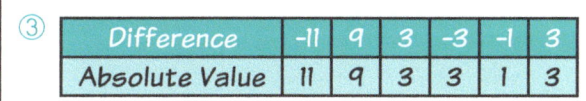

④ Mean of absolute values (mean absolute deviation):
$$\frac{11 + 9 + 3 + 3 + 1 + 3}{6} = \frac{30}{6} = 5$$

> The **mean absolute deviation** of Will's test scores is 5.
> This tells us that Will's data is fairly consistent.

A **mean absolute deviation** of 5 means that, on average, the spread of Will's test scores is 5 spaces away from the mean in either direction.

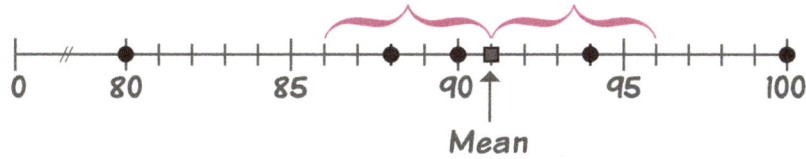

Topic Index

Absolute Deviation, 190
Absolute Value, 92, 145, 195
Acute Angles, 153–54
Addends, 25–26, 38, 44
   Unknown, 36, 41
Addition, 21–23, 25–26, 36–39, 42, 45–46, 50, 55, 125, 139, 157, 165
   Adding-On Method, 46
   Combining 100s, 10s, and 1s Strategy, 45
   Counting On Strategy, 44
   Decimals, 51–52
   Fact Families Strategy, 39
   Fractions, 108–110
   Making Ten Strategy, 44
   Traditional Method, 45
   Triangular Fact Cards and, 39
   Using Known Sums Strategy, 44
   Relationship With Subtraction, 21, 25, 27, 36–52, 139
   Sums, 38
   Whole Numbers, 44, 45
   Word Problems, 40–43
Adjacent Sides, 150, 152
Algebra, 25–26, 124–25, 133, 149
   Algebraic Expressions, 132–41
   Algebraic Thinking, 132, 183
   Tiles, 3, 139
   Vocabulary, 134
Algorithms, 4, 66, 73
   Push, 106
   Standard, 37, 45, 50, 55
   Traditional, 107
Angles, 150, 153–55, 157–59
   Equal, 157
   Markings, 155
   Measures, 153–54
   Opposite, 159
   Types, 150
Area, 55, 67, 115, 148, 164–65, 175–79, 181
Area Models, 62, 64–65, 67, 72, 79, 107, 115
   for Multiplication, 79
   for Whole Number Division, 72
   for Whole Number Multiplication, 62, 64, 65

Arrays, 10, 55, 59, 62, 64, 69, 80, 85–86, 111
Associative Property, 21, 25–26, 63
   of Addition, 25
   of Multiplication, 26, 63
Attributes, 148–50, 164
   Defining, 150
   Nondefining, 150
   Shapes, 150
Average, 191, 193, 195. *See also* Mean
Axis, 144, 145. *See also* X-Axis; Y-Axis

Bar Graphs, 183, 186, 188
Bases, 161, 177–78, 181, 183
Benchmark Numbers and Fractions, 29–30, 33–34, 103. *See also* Rounding
Box Plots, 183, 192–94

Categorical Data, 184
Coefficients, 134
Combinations
   of 10, 44
   Possible, 189
Common Denominator Method, 116–17
   Fractions, 116
Common Denominators, 104, 109, 116–17
Common Factors, 84
Commutative Property, 21, 26
   of Addition, 25
   of Multiplication, 26
Composite Numbers, 85–86
Composite Shapes, 163
Congruent Sides, 150, 152, 157
Conversion Factor, 127
Coordinate Grids, 144, 147
Coordinate Planes, 143–47. *See also* Coordinate Grids
Coordinates, 145
Cubes, 12, 87, 161–63, 180
Cups, 110–11, 125, 128, 170

Data, 182–95
   Analysis, 191, 193, 194
   Categorical, 184, 185, 186
   Measurement, 184, 187, 188
Decagons, 157

Decimals, 9–20, 32, 37, 51, 55, 67–68, 76, 79, 89, 95, 118–22. *See Fractions*
   Addition and Subtraction, 51
   Converting Fractions to, 119
   Division, 75, 76, 122
   Money, 121, 174
   Multiplication, 55
   Notation, 174
   Place Value, 8–19, 32
   Relationship With Fractions, 120
   and Percents, 120, 122
   Rounding, 32
Decomposing, 47, 49, 63, 99
   Addition, 44
   Fractions, 99
   Multiplication, 62–65
   Subtraction, 47, 49
   Whole Numbers, 99
Defining Attributes, 150
Denominators, 94–95, 97, 100–104, 108–9, 112–13, 117, 119–22. *See also* Common Denominators
   Addition/Subtraction, 108, 109
   Decimals, 120, 121
   Multiplication/Division, 112, 113, 116, 117
   Percents, 120, 122
Difference, 21, 37–38, 46–48, 134, 187, 192, 194–95
Distributive Property, 21, 26, 64, 136
Divide Fractions, 106–17
Dividends, 56, 74–76, 116
Division, 21, 27, 54–77, 111, 116–17, 122, 139
   Area Model Strategy, 72
   Array Strategy, 71
   Compare Strategy, 71
   Decimals, 76, 122
   Dividends, 56, 74, 75, 76, 116
   Divisors, 56, 75–76, 116
   Equal Groups Strategy, 71
   Fact Families Strategy, 57
   Fractions, 71, 77, 110, 116, 117, 122
   Long Division, 74
   Multi-Digit, 55, 66, 73
   Partial Quotient Method, 117
   Problems, 54, 56–57, 69, 76, 134
   Quotients, 56, 72, 73, 134
   Remainders, 75
   Strategies, 72–73
   Traditional Method, 72, 74
   Triangle Facts, 57
   Word Problems, 58, 69
Divisors, 56, 75–76, 116
Dot Plots, 183, 187

Edges, 5, 87, 161–62
Equal Signs, 23–24, 135
Equations, 23, 56, 77, 132–41, 146
Equilateral Triangles, 158
Equivalent Fractions, 3, 34, 95, 101, 104, 112–14, 122
Equivalent Ratios, 127–30. *See also* Ratios
Estimation, 28–34, 67
   Front-End, 29, 33
   Rounding, 30–32, 34
   Strategies, 29, 33
Even Numbers, 80
Exponents, 27, 79, 87
Expressions, 24, 134–36, 138, 141

Faces, 161–62, 179
Fact Families, 39
Factors, 56, 63, 68, 79, 83–86, 134, 140
   Common, 84
   Friendly, 73
   Greatest Common, 84
   Shared, 84
   Unique, 86
   Unknown, 54
Fluid Ounces, 170
Fourths, 95–96, 102, 118. *See also* Quarters
Fractions, 16–18, 51, 68, 94–104, 106, 176
   Addition, 107
   Common Denominators, 104, 108–9
   Comparing, 102–3
   Converting to Decimals and Percents, 122
   Decimal, 121
   Denominators, 97–98, 100, 102, 108–9, 121
   Division, 71, 75, 101, 107, 116, 117
   Equivalent, 101, 104, 108, 109
   Estimation, 34. *See also* Benchmark Numbers and Fractions
   Fraction Bars, 120
   Fraction Circles, 108, 172
   Mixed Numbers, 109, 110, 114

Multiplication, 101, 107, 112, 113, 114, 115
Numerators, 97–98, 100–2, 108, 112, 113, 117
Reciprocal Multiplication Division Method, 117
Subtraction of, 108–10
Unit, 98
Whole Numbers, 100, 112
Word Problems, 110, 111
Friendly Numbers, 30, 46–48. *See also* Benchmark Number
Functions, 125, 133, 143, 146

Gallons, 110, 125, 170, 193
Geometry, 96, 148–63, 183
   Graphing, 144–147
   Line, 151
   Line Segment, 151
   Point, 151
   Ray, 151
   Vocabulary, 151
Grams, 171
Graphing, 146, 181
   Equations, 143
   Lines, 143
   Numerical Patterns, 147
   Word Problems, 146
Graphs
   Bar Graph, 186
   Dot Plot, 187
   Line Plot, 187
   Picture Graph, 185
Greater Than. *See* Inequalities
Greatest Common Factor, 84

Halves, 95–96, 113
Hexagons, 157
Histograms, 183, 188
Hundreds, 9, 11–13, 15, 18–19, 45, 49–50, 52, 63, 66, 72, 74
Hundred Charts, 81
Hundred-Thousands, 13
Hundredths, 9, 16–18, 20, 32, 67, 95, 119–22

Identity Property, 21, 26
   Addition, 25
   Multiplication, 26
Independent Variables, 140, 146
Inequalities, 132–41

Integers, 3, 82, 88–92, 120, 134
   Negative, 89, 143
Interquartile Range, 190, 192, 194
Inverse Operations, 36, 39, 54, 57, 139
Inverse Relationship, 36–37, 54
Isosceles Triangle, 158

Kites, 160

Least Common Multiples, 82, 104
Less Than Symbol, 24. *See also* Inequalities
Line of Symmetry, 152
Line Plots, 187
Linking Cubes, 2, 56, 60, 179, 181
   and Nets, 179
Liquid Volume, 165, 170–71
Liters, 171
Long Division, 74
Lower Extreme, 193–94
Lower Quartile, 193–94

Math Symbols, 2, 22
   Equal Sign, 22, 23
Mean, 56, 183, 190–92, 195
Mean Absolute Deviation, 192, 195
Measurement, 155, 164–81, 183–84, 190
   Angles, 154–55
   Area, 175–78
   Data, 183–84
   Length, 145, 166–69, 171
   Liquid Volume, 170–1
   on a Coordinate Plane, 145
   Perimeter, 175, 176
   Volume, 180, 181
   Weight, 169, 171
Measurement Data, 186–7
Measurement Units, 127, 169–71
Measures of Center, 190–92
Measures of Variation, 190, 192
Median, 183, 190–91, 193–94
Meters, 128, 171
Metric Units, 165
Millions, 13
Mixed Numbers, 70–71, 109, 114
Mode, 183, 187, 190–91
Money, 16, 121, 138, 165, 174, 181
Multi-Digit Numbers, 11, 37
   Multiplication Strategies, 55, 64–65
   Place Value, 11

Multiples, 61, 63, 79, 81–82, 100, 104, 137, 147, 171
    Common, 82
    Least Common, 82, 104
Multiplication, 21, 26, 54–60, 62–64, 67, 69, 79, 111, 139, 165, 189
    Area Model Strategy, 62, 64, 65
    Array Strategy, 62, 64
    Associative Property, 63
    Decimals, 67–68
    Decomposing Strategy, 62–65
    Distributive Property, 65, 136
    Division, Relationship With, 27, 54–77, 139
    Equal Groups Strategy, 59, 62, 69, 111
    Exponents, 87
    Fact Family Strategy, 57
    Fractions, 112–15
    Multi-Digit, 55
    Multiples, 63
    Partial Product Strategy, 64, 65
    Picture Method Strategy, 112–114
    Repeated, 87
    Scaling Factor, 59–60
    Skip Counting Strategy, 62–64
    Traditional Method, 112–14
    Triangle Facts Strategy, 57
    Word Problems, 58, 59

Negative Numbers, 88, 90–91
Nets, 149, 161–62, 179
Non-Defining Attributes, 150
Number Lines, 2, 5, 30, 32, 46, 48, 89, 91–92, 94–95, 97, 106–7, 187–88, 193–94
    Double, 128
    Half, 89
    Horizontal, 91, 92, 97
    Open, 30
    Vertical, 91
Numerators, 94, 97–98, 100–103, 108, 112–13, 117

Obtuse Angles, 153–55
Octagons, 157
Odd Numbers, 80–81
Ones, 10–15, 19
Ordered Pairs, 145–46
Order of Operations, 27, 77, 136
Origins, 144
Outliers, 193–94

Parallel, 92, 151, 157, 159, 161
    Lines, 151
    and Opposite Angles, 159
    Sides, 150, 157, 160
Parallelograms, 157, 159–60, 177–78
    Special, 159
Partial Quotient Method, 72–73
Patterns, 9, 61, 79, 81, 137, 147
Pentagons, 94, 157
Percentiles, 194
Percents, 118–22. *See also* Fractions
Perimeter, 127, 148, 165, 175–76
Perpendicular Lines, 151
Pictographs, 185
Picture Graphs, 183, 185
Pints, 170
Place Value, 3, 8, 11, 17, 19, 31, 36–77
    Decimals, 15–20, 32
    Disks, 12, 50
    Organization, 14–15
    Whole Numbers, 11, 12, 13, 19, 31
Polygons, 149, 156, 175–76
Positive Numbers, 88–91, 195
Prime Numbers, 85–86
Product, 26, 56, 59–65, 67–68, 72, 82–83, 109, 112, 115, 134–35, 137
    First, 66
    Partial, 64–65, 73, 115
Product Factors, 56
Product Unknown, 58
Properties, 21–27, 63, 79, 148–49
    Geometric, 149
Properties
    of Addition, 25
    of Multiplication, 26
Protractors, 155

Quadrants, 143–45
Quadrilaterals, 157, 159–60
Quarters, 96, 169, 172, 174
Quartile, 193–94
Quarts, 170
Quotient, 56, 74, 76, 116, 134

Range, 187, 190, 192
Rates, 55, 124–30, 146. *See also* Ratios
    Rate Table, 130
Ratios, 95, 120, 124–27. *See also* Fractions
    Convert Measurement Units, 127
    Equivalent Ratios, 128–30

    in Fraction Form, 120
    Part-To-Whole, 94
    Ratio Table, 129
Rational Numbers, 120
Rectangles, 67, 96, 115, 157, 159–60, 163, 175–77, 179
Remainders, 75
Rhombuses, 157, 159–60
Right Angles, 151, 153–54
Right Circular Cone, 163
Right Circular Cylinder, 163
Right Rectangular Prism, 163
Right Rectangular Pyramid, 163
Rounding, 28–30, 32–33. *See also* Estimation
    Decimals, 32
    Large Numbers, 31
    Rules, 30
    Remainders, 75

Scalene Triangles, 158
Scaling Factor, 59–60
Shape Attributes, 150
    Adjacent Sides, 150, 152
    Angle Types, 150, 153–54, 159
    Congruent Sides, 150, 151, 157
    Number of Angles, 150, 157–59
    Number of Sides, 150, 157–59
    Parallel Sides, 150–151
    Perpendicular Sides, 150–51
    Symmetry, 150, 152
    Vertices, 150, 157
Shapes, 2, 94, 96, 143, 148–50, 153, 156–57, 159, 161–63, 175, 178, 180, 190
    2D, 157, 163
    3D, 149, 161
    Attributes, 149
    Break, 148
    Composite, 163
    Flat, 161
    Four-Sided, 160
    New, 148
    Partition, 95
    Solid, 161
Skip Counting, 62–64
Squares, 10, 52, 87, 94, 106, 117, 157, 159–60, 163
Statistical Question, 190
Statistical Variability, 190
Straight Angles, 154

Subtraction, 21, 27, 36–52, 107, 110, 139
    Adding-On Strategy, 46
    Compensation Strategy, 48
    Decimals, 51–52
    Decomposing Strategy, 49
    Fact Families Strategy, 39
    Fractions, 108
    Mixed Numbers, 109
    Problems, 36, 38–39, 134
    Regrouping, 50
    Relationship With Addition, 39
    Shifting the Number Line Strategy, 48
    Traditional Strategy, 50
    Triangular Fact Cards, 39
    Whole Numbers, 47–50
    Word Problems, 40–43
Sums, 25–26, 38, 44, 46, 99, 134, 141, 147
Symbols, 2–3, 19, 21–22, 27, 37, 56, 94, 132, 134, 153, 185
Symmetry, 149–50, 152

Tape Diagrams, 128
T-Charts, 137
Tens, 10–12, 19
Ten-Frame Tiles, 12, 60, 66, 73, 139
Ten-Thousands, 13
Tenths, 16–18, 20, 51, 67, 119–21
Terms, 2, 60, 134–35
Thirds, 95–96, 112–14, 172
Thousands, 12–13
Thousandths, 16–18, 20, 119–20
Time, 172–3, 178
Total, 38, 41. *See also* Sum
Trapezoids, 157, 160
Triangles, 39, 57, 150, 157–58, 163, 177–79
    Equal-Sized, 94
    Faces, 179
Triangle Facts, 39, 57

Unit Fractions, 5, 98
Unitizing, 10
Unknown Numbers. *See* Variables
Unknown Angles, 155
Unlimited Number, 127
Upper Extreme, 193
Upper Quartile, 193–94
U.S. Standard Measurement Units, 169–70

Variability, 183, 190, 192, 195
Variables, 26, 40–41, 58, 133–36, 139–40, 143
   Independent and Dependent, 146
Vertices, 150, 157, 161–62
Volume, 6, 148, 161, 164–65, 180–81
   Liquid, 170–71

Whole Numbers, 11–13. *See also* Integers
   Addition, 44, 45
   as Fractions, 90, 100, 101, 112, 116
   Division, 72–75, 77, 116
   Multiplication, 61–66
   Subtraction, 46–50

Word Problems, 37, 40–41, 66, 132, 135, 138, 165, 167–68
   Addition, 40-43, 77, 110
   Division, 69, 77, 111
   Fractions, 110
   Measurement, 167
   Subtraction, 40–43, 77, 110
   Two-Step, 43, 77
Writing Equations, 58, 138
Writing Expressions, 135, 138
   Solving, 110

X-Axis, 144–45

Y-Axis, 144–45

CORWIN Mathematics

Supporting TEACHERS | Empowering STUDENTS

PETER LILJEDAHL

Fourteen optimal practices for thinking that create an ideal setting for deep mathematics learning to occur.

Grades K–12

GRADES 6–12 COMING SOON!

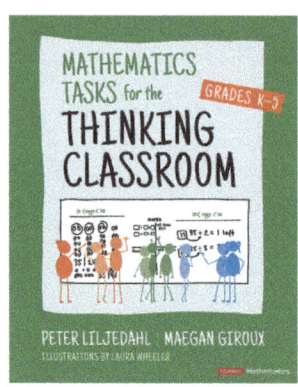

PETER LILJEDAHL, MAEGAN GIROUX

Delve deeper into the implementation of the fourteen practices from *Building Thinking Classrooms in Mathematics* by focusing on the practice through the lens of tasks.

Grades K–5

RACHEL LAMBERT

Discover UDL for math, a way to design math classrooms that equips all students for meaningful and joyful math learning.

Grades K–8

VANESSA VAKHARIA "THE MATH GURU"

Equip students to develop the skills they need to truly believe anything is possible, even a better relationship with math!

Grades K–12

STEVEN LEINWAND, CAROLINE WELTY

Create a positive learning environment for your students and help them boost self-confidence, reduce math anxiety, and master essential skills.

Grades K–12

KRISTOPHER J. CHILDS, JOHN W. STALEY

Make mathematics learning relevant and useful for students as they use mathematics to understand the world around them and celebrate their unique identities.

Grades K–12

To order, visit corwin.com/math

HOLLY BURWELL, SUE CHAPMAN

Build a vibrant math community and maximize your students' math learning with this hands-on, 10-month practice-based professional learning guide.

Grades K–5

PAMELA WEBER HARRIS

Guide students through domains of math development, from counting and adding to more complex proportional and functional reasoning—*without* resorting to algorithms.

Grades K–12

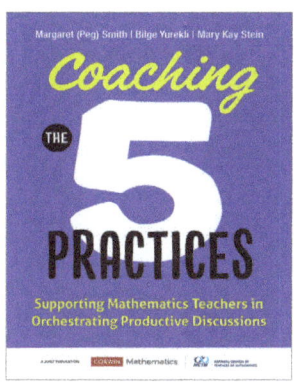

MARGARET (PEG) SMITH, BILGE YUREKLI, MARY KAY STEIN

Refine your craft and become increasingly more skilled at implementing Smith and Stein's renowned 5 practices in ways that support the learning of each and every student.

Grades K–12

JO BOALER, CATHY WILLIAMS

Introduce data science to your curriculum across disciplines with real-world stories and teacher testimonials to transform your classroom experience.

Grades K–12

BERTHA VAZQUEZ, KIMI WAITE, LAUREN MADDEN

Use this inspiring road map to integrate climate change lessons into your existing curriculum and foster student agency across disciplines.

Grades K–12

IRINA LYUBLINSKAYA, XIAOXUE DU

Integrate AI literacy into K–12 classrooms, blending theory, practical lesson plans, and ethical considerations to empower students as critical thinkers.

Grades K–12

CORWIN

To help every educator help every student

We believe that every single student deserves a great education

We believe that knowing our impact is both a privilege and a responsibility

We believe that a fair, stable, and thriving society is built on education